Evolution in Color

Frans Gerritsen

Schiffer Publishing Ltd

1469 Morstein Road, West Chester, Pennsylvania 19380

For Ruthie

This work was originally published in Dutch as *Evolutie van de Kleurenleer* by Cantecleer bv, de Bilt, 1982. Copyright © 1982 Cantecleer bv.
Translation by Dr. Edward Force and Ruth de Vriendt
Didactic illustrations by Frans Gerritsen
Reproduction drawings by Nico Raemaekers

English edition copyright © 1988 by Schiffer Publishing Ltd.
Library of Congress Catalog Number: 88-61472.

All rights reserved. No part of this work may be reproduced or used in any forms or by any means—graphic, electronic or mechanical, including photocopying or information storage and retrieval systems—without written permission from the copyright holder.

Printed in the United States of America.
ISBN: 0-88740-143-0
Published by Schiffer Publishing Ltd.
1469 Morstein Road, West Chester, Pennsylvania 19380

This book may be purchased from the publisher.
Please include $2.00 postage.
Try your bookstore first.

Contents

1 Evolution in color 4
2 Colors and Emotional Values 5
3 The Visual Color Image 7
4 The Main Development Line of Color Theory in Color Ordering ... 9
5 Linear Color Ordering 10
6 The Colors of the Daylight Spectrum 11
7 The Circle as a Basic Form 13
8 Main Divisions of the Basic Colors 14
9 The Unique Primeval Colors in a Color Ring 15
10 Mixing Properties of Paint Colors 16
11 Color Perception as a Basis 49
12 From Full Color Hue to White—From Full Color Hue to Black 51
13 All Intermediate Colors From Full Color Hue to White, to Black and to All Gray Values 51
14 Distances From Full Color to White and Black are Equal 52
15 Lightest Colors Closer to White—Darkest Colors Closer to Black 52
16 From One Lightest to One Darkest Color 53
17 Three Dark Basic Colors—Three Light Basic Colors 54
18 The Color-Perception-Space, Gerritsen—1975 55
19 Color Perception and Color Hue 58
20 Color Perception and Lightness 61
21 Color Perception and Saturation (Colorfulness) 62
22 The Three-Dimensional Schematic Color Perception Diagram ... 63
23 Color Cross-Sections of the Color Perception Space 64
24 Evaluation of Specialized Literature 81
25 Build Your Own Model of the Color Perception Space 82
26 Emotional Color Values 84
27 Professional Pointers 86

1 Evolution in Color

Evolution

The development of ideas about colors

Evolution is the development from an outdated or past stage to the next phase of development. It is a process which will always be repeated until, perhaps, the end of time. Color theory is the study of colors, the colors of the sky and the rainbow, the colors of the oceans and the lakes, the colors of plants, flowers and fruits, the colors in the world of man and animals. We think of colored leaves in green, yellow, orange, red, brown and violet. We see flowers in the colors of the spectrum, in magenta and brown, in various light pastel colors, and flowers in splendidly glowing and dark colors; the same is true for the color of fruits. We think of peoples with different skin colors, and with different eye and hair colors. In the animal kingdom there are colorful tropical fish, bright-colored beetles and butterflies; birds such as the humming bird, pheasant, peacock and parrot; our house pets such as dogs, cats, marmots, rabbits; and all the other different kinds of animals such as ermines, foxes, tigers, giraffes, zebras...

Colors in nature

Colors in art

There are the colors that man has made, the colors of paint on the painters' palette or the oil painting, printing inks for color printing, the colors of photo and film, the colors of colorful costumes and other textile (art)forms. Think of colored glass, stained glass windows, colorful engobes, colorful glazes and enamels. Think of the varicolored images projected in stage, flood and spot lighting for theater and ballet; mercury, sodium and neon lights, fluorescent colors and phosphor 'spots' as activated in color television.

Colors and utility

The observation and study of the many forms in which colors appear, and the insights thereby gained, have led to theories of the many characteristics of color, the 'theory of color'.

A clear line of development

A number of different aspects of color have been presented systematically in my book: Theory and Practice of Color, (published in seven countries). This publication, Evolution in Color, traces a clear line of development, through that mass of information, in the ordering of colors from the Sixth Century B.C. to the present. We discuss this ordering of colors which we observe as part of our emotional experience in the colorfulness of our everyday life.

Colors from light and darkness

If one follows the clear line of development in search of the 'nature' of colors, one first encounters the assumption of the ancient Greeks that colors originate from the struggle between 'light' and 'darkness', between white and black.

Colors and their wavelengths

Much later, after the Middle Ages, it was believed that the existence of colors could be explained by dividing white light into light of various wavelengths, using a glass prism.

Colors originate through 'sight'

Today we believe that colors originate within ourselves, through our visual organ, through our eyes, our brains and our consciousness.

The light and dark values of colors

Color contrasts, but also light-dark contrast values, are essential for the artist in terms of the 'legibility' of his work. For that reason we have put particular emphasis on the question of how light or dark a color is in the diverse color orders. This concerns, above all, the locations of the colors, in each successive color order (theory or diagram), as regards their light-dark values, as compared with black and white.

Illustrating the color ordering systems with colored dots

Colored dots

In the diagrams of the various color ordering systems (pp. 17-48), the given place for each colored dot is marked with a small circle and a number. On pages 79 and 80 you will find small squares of color which are likewise numbered. From this page you can cut small circles of color by using a 10 mm punch. The numbers in the circles of the diagrams must match those of the color samples to be applied (see pp. 17-18).

Model of the color perception diagram.

This book also includes instructions for making your own color-perception-space model. It also can aid the reader in better use of specialized literature and professional guidelines.

The next chapter, we shall first turn our attention to the colors and their emotional values as we experience them in our everyday world.

2 Colors and Emotional Values

Visual values —emotional values

We must first consider the real value of color: the emotional value of color in our daily lives, before looking into the different theories which have been thought out to explain (ordering of) visual color factors.

A color as 'color' perception

When we talk about white, everyone will know generally what we mean. White is the opposite of black, does not incline toward any color, is neutral and has the highest degree of lightness. It is not easy to define white, but we all know what is meant when the color of a surface is described as white.

Color and emotion

However, when color affects us emotionally... We wake up in the morning and unexpectedly we see that our whole world is white!... It has snowed. Everything outdoors is suddenly enchantingly beautiful. All is very still, and the white world looks immeasurably vast. The charm of such scenery can make us feel that life is worth while, and that there are many of these highlights such as: the sparkling stars on a winter's night, broad untouched beaches and the eternal rhythm of waves, the deep blue of the sky above a field of waving golden wheat in the summer, the thin air which seems to lose itself in the magnificent fjords, the blazing colors of fall and the rich darkness of freshly plowed fields. The white that causes us to daydream is very different from the white of a white colored surface as a contrast to a black colored surface.

Positive and negative influences

Naturally, the world of color experience is not always charming. We can imagine a gray surface and remain rather indifferent to it. The same leaden gray in the sky, just before a thunder storm breaks, however, can be experienced as threatening, as a representation of evil demons. This kind of experience causes us to realize that it is not always a matter of course that we are safe. Suddenly we realize that we are in a world full of catastrophic possibilities: floods, earthquakes, war and hunger, violence and hate.

Situation-association and emotional values

We know, from experience, that white of the snow, seen in a different situation, can mean the hopeless end after interminable waiting for the life-saving thaw. The dark gray sky can also signal the long-awaited rain after a terrible draught. We see here that different situations can cause very different emotional reactions to exactly the same white and precisely the same gray.

Generally accepted color values

Warm and cold colors

There are also general emotional color associations. There are warm and cold colors, hard and soft colors, businesslike, romantic, happy and somber colors. Magenta, red, orange and yellow are regarded as warm colors; blue, blue-violet and blue-green, as cold colors.

Deviations from generally accepted color values

Considerably different emotional color values can be present under special circumstances. Welders and enamelers know that the blue gas flame is much hotter than the yellow, orange or red colored flame. Temperature radiators, when measured in degrees of Kelvin, also show that the greatest heat lies in the blue, and the lower temperatures are in the red area of the spectrum. Even so, we call blue light 'cool' and red light 'warm'.

Colors as primal powers

We all know the natural process of burning, but we are always overwhelmed by the angry sea of fire of a burning oil rig, a prairie or a large metropolitan fire. Therefore, we can understand how man, in earliest times, thought of colors as the personification of primeval forces:

Subconscious color associations

Red: the roaring force of fire, blood, war and violence, symbolic color of the ego.
Green: the primal power of fertility which makes seed sprout, force of peace and prosperity, symbolic color of rest.
Blue: the might of the infinite, the heavens and the firmament, of thoughts and meditation, symbolic color of space and eternity.
Yellow: the power of the laws of chronos, of time, the sun, moon and stars, the mightiest upper-god, symbolic color for 'shining', for time duration and 'temporality'.
Black: the sinister power of darkness, death and decay, symbol of mourning.
White: the blinding 'light' of the spirit, triumphant over death, the white ashes after the fire, the silence of snow that covers the black winter wood, symbol of purity, the untouched.

Concepts linked with particular colors

Other conceptions which were held in olden times regarding certain colors are: 'the elements of which everything is made', air-earth-fire-water; the four quarters of the compass: north, south, east and west; the seasons: spring-summer-autumn-winter; concepts of the struggle for survival; those from religion and heraldry; philosophies of anthroposophists and finally, the traditional symbol-colors of the national and internationally normalized form and color symbols. Many primal color associations, conscious as well as unconscious, still influence our emotional reactions to color in the world around us. These emotional values are quite different, from one individual to another, because of 'personal' experience associations with certain colors. There are times when we can enjoy life in full measure, and other times when our world seems to be disintegrating. When a certain color evokes an association with such an experience, it is possible that we can no longer react objectively. Years later, such a color can, unconsciously, be experienced as pleasant, or perhaps in another situation seem disagreeable and unacceptable.

Individual influences

Common group influences

The more universal emotional values, of the various colors, are bound by time, place, national traditions, development stage, age, sex and fashion.

The visual color image

All the different possibilities of emotional color value per color are one and the same color perception. Evolution in Color will trace the ordering of visually perceived colors from 600 B.C. to the present. The artist gives his message to the world through emotional values of colors and through the interplay of surface relationships and forms. He evokes these emotional color values by means of visually unequivocal colors into which he has 'translated' emotional values.

Emotional values 'translated' into unequivocal colors

3 The Visual Color Image

Colors are attuned to daylight

When speaking of colors which we can perceive, the visual color image, we must agree upon color names and descriptions. Only then can we compare the various color orderings with each other. The colors of the illustrations are attuned to normal daylight, therefore they are best viewed by daylight.

Color-descriptive concepts

We will discuss the following color-descriptive concepts: the color names; the unique colors; the color hue; the lightness▫brightness of the color; the saturation of the color; mixing principles of colors: subtractive, additive and partitive.

The color names

The color names, or abbreviations, in the illustrated color diagrams, have, whenever possible, been given in the native language of the individual author/color scientist. Unfortunately, there is no uniformity in the naming of colors, so that a certain color is often listed by different names in the same language, (#2, p. 19).

The unique colors

We will use a negative selection, for the sake of clarity, as to the essence of several main colors and their names. This manner of color description is attributed to Leonardo da Vinci. It forms the basis of the Scandinavian Natural Color System (NCS; #28 & 68).

Anders Hård, from Sweden, again emphasized the description of these unique colors at the 1969 AIC International Color Congress in Stockholm. It is possible to describe any given color by means of only six unique color concepts, (#3, p. 20).

The three dimensions of a color

1. The color hue

The color of a color

The color hue of a color is the concept that the color of a color designates; for example: the color hues blue, yellow, green, red etc. The words 'color hue' and 'color tone', (in french: tonalit (chromatique), teinte; in german: Farbton), are derived from the concept of 'tone' in music, which defines sound. The colors which lie in between: orange, orange-red, blue-green, etc., are also color hues. A closed color ring is made up of a succession of the infinite number of color hues, (#4, 117).

2. Lightness

How light or dark is a color?

We use the word lightness to describe how light or how dark a color is. Lightness is a succession of lightness steps from white to black. Outdated terms for these same concepts are: brightness, grayvalue. The word 'brightness' is used to describe the brightness perception of light sources, as in bright or dim lights, (#5).

3. Saturation

The intensity (chroma) of a color

Saturation indicates the colorfulness of a color, the amount of full color hue as opposed to the huelessness of a color.

White, the gradations of gray and black are, to color scientists, known as colors without color hue or colorless colors. They are also called 'achromatic colors' to clarify (and sometimes confuse) our language. Painters use the words: neutral, achromatic, or colorless colors. Psychologists also make a distinction between colors and non-colors. Ostwald called them, wrongly, 'bunt' and 'unbunt', which means multi-colored and nonmulti-colored colors.

Colors and non-colors

One solid color cannot be multi-colored; however, speech concepts change easily, so words such as 'bunt/unbunt' might well be coined to describe an idea for which an acceptable word does not yet exist.

There is no proper differentiation of words for this concept in our languages. If we speak of the properties of an object, which we call color, we can say: the color of that object is white, yellow, gray, red, green, black or blue. If, however, we wish to discuss concepts of the perception of colors or non-colors, we can differentiate between a color and a black and white photograph—which has no color, only differences in lightness. All this is necessary because our language does not provide us with possibilities to name a colorful color and a colorless color as different parts of the same concept, (#6).

Our language is deficient

Mixing principles of colors

Our eyes translate, together with our brain, light of different wavelengths (singly or in combination) into different colors. There are three mixing principles: subtractive, additive and partitive.

Subtractive

Subtraction of light

We subtract part of the light in the direction of the eye in subtractive mixing (transparent color layers over each other), (#139, 140 & 141).

Practical applications: water-colors, enamels, glazes, glass collages, stained glass, layered filters, color slides, color film and photos.

Additive

Addition of light

The addition of light in the direction of the eye through mixing, is called additive color formation (colored spotlight beams projected one over the other), (#142 & 143).

Practical applications: ballet and theater stage lighting, floodlights, the color image of a television screen with the glowing, phosphorescing spots of blue, green and red (the spot pattern remains partitive).

Partitive

Averaging of light

The averaging (the average), of a part of the light in the direction of the eye through mixing, is called partitive color formation (a pattern of very small color surfaces, color spots, next to each other) (#144, 145 & 146).

Practical applications: the spots of three-color printing, the dot pattern of a television screen, the small colored stones of a mosaic, multi-colored woven patterns.

Light-dark color values and their ordering

Light-dark contrast

We know, from practical experience, that a light/dark contrast is 'easier to read' than a large color contrast between colors of equal lightness. Color compositions, or completed color schemes, will show more light/dark than color contrast, with equal light/dark values.

Color contrast

Light/dark values, or contrasts, are also considered very important in painting. The light/dark values, consisting of light, medium and dark sections, are described as the 'tone' of a painting. The total impression of a painting, constituting many colors, is often described as sunny or gloomy of tone.

Difference between color hue and tonality, 'tone'.

The internationally accepted nomenclature, (CIE nr. 17, 1970-1977), uses the words 'color hue' to refer to the actual color: red, blue, green, yellow, or a mixture of two neighboring colors in the color circle. The word 'hue' can easily lead to misunderstanding. Whenever possible, we have, in this book, used the most up to date internationally accepted nomenclature for the naming of color concepts. The light/dark values of colors greatly determine the visual form/image. This is why we will place extra emphasis on the place which light and dark colors will occupy in color ordering.

4 The Main Development Line of Color Theory in Color Ordering

The development of color theory clearly shows the three different approaches that have been used for color ordering.
The development of color theory clearly shows the three different approaches that have been used for color ordering.

1. Colors originated from the struggle between darkness and light; a color ordering from light to dark, from white to black, evolved from this area.
2. Colors originated by the dissention of white daylight into light of a continuous series of different wavelengths.
3. The colors originate through sight and are therefore linked to the color perception possibilities of our visual organ.

Philosophies about the origin of colors from 600 B.C., the ancient Greeks

Aristotle

The Greeks observed color behavior and felt that colors had a primal power which caused emotional reactions; this was reflected in their use of color symbolism. Aristotle tried to examine the mixing characteristics of colored light. He let daylight shine through a small piece of yellow glass onto a white marble surface and saw that the colorless light had become yellow. He tried the same experiment with a blue piece of glass and saw a blue light spot on the marble. When he let the light pass first through the yellow and then through the blue glass, a green patch of light appeared on the marble surface. When the colored glasses were shifted a bit, he saw the colors yellow, green and blue next to each other. He drew the wrong conclusion from this experiment: that yellow light mixed with blue light produces green light.

A false conclusion

If he had projected the yellow light and blue light separately, with a mirror, onto one spot on the marble surface, he would have found that mixing yellow and blue light produces white light and not green (#143).

Yellow and blue light together white

What did Aristotle do? He let the yellow light pass through the blue-colored glass. The part of the yellow light, which was passed through the blue glass, was only the remnant of the yellow, and this remaining light looked green. This is not a mixture of yellow and blue light, which produces white, but a filtering of yellow light through a blue piece of glass, which also lets green through. That is a totally different process of color formation. The same wrong inference, naturally, was made when the succession of colors in the rainbow was studied.

The origination of colors

The origination/formation of colors was sought while studies of color properties and the mixing results, etc. were made. It was believed that the shadows behind the raindrops in the clouds made colors come into being. It was thought that the colors of the ocean and in other colored objects were similarly formed, by the shadows behind small particles of water and behind small particles of solid materials. Another conception was that colors came into being through the struggle between light and darkness, between white and black.

The struggle between light and darkness

When white light has passed through a sizeable cloud, the white has become darker. The cloud no longer looks white, but gray, sometimes dark gray. The sun, in the early morning, is red. As soon as the sun climbs higher, the light has a shorter path to follow through the diaphane, or transparence, the red color becomes lighter. The higher the sun rises, the shorter the path that the light must travel through the diaphane and the lighter it becomes. Red changes via orange and yellow to white. When the sun reaches its zenith, at midday, light has the shortest path to follow through the transparence; the light is white. The whole process is repeated in reverse order as evening approaches, as the light path through the diaphane (atmosphere) becomes longer, colors change again. The white light first becomes yellowish and then changes via orange back to red. When the sun has set, the light has an even longer way to travel. Now the red of sunset changes via purple-violet into the deep blue night firmament. Finally night falls and the color of the sky changes to black.

Red as a mixed color between white and black

Sometimes color formation could be noted in objects. A black steel mirror spotlighted by white light, appeared to have a reddish glow. The mixed color between white and black was red here, too, just as with the sun. There, the red color of the rising and setting sun lay between the white of the day and the black of the night.

If we almost close our eyes so that very little light passes through our eyelashes, all kinds of colors appear. This phenomenon was also seen as proof that the longer or shorter path that the light must travel, here because of the obstruction by the eyelashes, is the cause of the color formation.

5 Linear Color Ordering

A linear color ordering originated from the concept that colors are formed through the struggle between light and darkness; herein the colors are arranged from white as the lightest color to black as the darkest color (#7 & 8). Red has its place midway between white and black, this according to its own lightness in comparison to the lightness of white and the absence of lightness in black. Yellow or gold lies between white and red, darker than white but lighter than red. Ultra-marine blue is placed past red,

Colors ordered according to their inherent lightness between black and white

it is darker than red but lighter than black. Green is incorrectly seen as the mixed color of yellow and blue light and is placed together with red, midway between white and black (#8).

This color ordering, according to the relative lightness of the diverse colors themselves, lasted for about 2200 years, from 600 B.C. till about A.D. 1600.

Della Porta

The colors of the rainbow

The color diagram of Della Porta, 1593 (#9), shows the refraction of sunlight through a prism. He explains, in color diagram #10, the rainbow colors in a rain cloud. The area L-E-D-I represents the rain cloud, point A is the location of the sun, and B is the location of the observer. The refraction of the sunrays is given in points F. P. O and M. The F-P pathway is yellow, according to Della Porta, the O-M path is blue; the P-O path, in between, is green because of the mixing of the blue and yellow paths. This is still the result of Aristotle's mistaken thinking; he believed that the mixed color of yellow and blue light was green. Della Porta does not see the red as originating through refraction, but through mixing white light with the black of the cloud.

Aguilonius

The primal colors and mixed colors

Aguilonius' color diagram, 1613 (#11), shows not only the linear arrangement from white to black, but also the different mixing possibilities, which are indicated by arcs.

Kircher

2200 years of color ordering from light to dark

Kircher's diagram, 1646 (#12), incorporates a more consistent conception of mixed colors, indicated by semicircles. This marks the end of a period in which countless color researchers, over a period of 2200 years, ordered the colors in a straight line from the lightest to the darkest color, beginning with white and ending with black.

It is not until the Twentieth Century that the lightness of the basic colors once more gets attention colors were ordered as to their own intrinsic lightness, according to the lightness scale from white to black.

6 The Colors of the Daylight Spectrum

Colors were thought to originate through the division of white daylight into light in a series of successively increasing wavelengths.

Color ordering from light to dark disappears

A color ordering based on the succession of colors in the daylight spectrum is created. The ordering of colors from light to dark, in relationship to white and black, which is so essential to the artist and designer, was abandoned after more than 2200 years. Complementary color pairs do not lie opposite to each other in this phase of color ordering. The color magenta is missing from this color series.

The old color ordering from light to dark, from white to black, which existed from 600 B.C., was breached in the Seventeenth Century by Sir Isaac Newton's color ordering.

Many reasons can be named which obstructed new theories during the 2200 years between 600 B.C. and Newton. There was the lack of easy communication because of relatively long distances. The church and state dogmas restricted thought, (cf. Galileo Galilei, 1546-1642). There was little knowledge of physics and physiology, and a limited choice of color pigments. Artists and painters, who prepared their own paints, tended to keep the basic ingredients and mixing ratios strictly secret, like magical formulas.

Newton dissects daylight through a prism

Sir Isaac Newton, 1642-1725, breached the accepted color ordering from white to black. He dissected white daylight, with the help of a glass prism, into light with a fanned-out band of successive wavelengths (#90, above the black line with an arrow). The color hues which we can perceive here are: violet-blue, ultra-marine blue, cyan blue, green, yellow, orange and red.* Newton named these colors: violet, indigo, blue, green, yellow, orange and red. He placed these color hues into a closed circular color ring, in which violet-blue and red came together. He placed white within the enclosed space of the color ring, to symbolize the combining of all light spectrum colors to white (#13 & 14). White and Black are no longer included in this new circular arrangement of the colors.

The color ring

Newton's color circle did away with the centuries-old color ordering of the light/dark values, the lightness values and the color hues in relation to white and black.

Ordering as to the intrinsic lightness of colors is lost, this loss is not noticed

The lack of the light/dark relationship in the ordering of color hues was a significant loss for (color)artists. This is because the light/dark contrast has a greater visual contrast effect than a color contrast of equal lightness. It is, therefore, important to know how light or how dark a color hue is. This loss, which occurred in Newton's color ordering, was hardly noticed and certainly never analyzed.

Fierce reactions against the 'new'

It was the contradiction of age-old color laws which invited fierce reactions against Newton's work; these reactions came from others who also studied colors.

A historical example of this negative criticism is found in Goethe's 'Color Theory', more than a hundred years later:

White Newton made from all colors.
He deceived you so well that you believed him a century long.
*
Go ahead, split the light! You try to separate, as you often have,
that which is one and remains one in spite of you.
*
A great physicist teaches us, together with his followers:
'Nil luce obscurius!' (nothing is more obscure than light)—
indeed! for obscurants.
*
Newton erred?—yes, many times over.
How then?
Printed long ago, but read by no one.
*
A will-of-the-wisp I saw disappearing, you Phlogiston!
Soon, O Newtonian shadow, you'll follow your little brother.
*
Newton observed reasonably but concluded wrongly.
British to the end: persistently he drew conclusions,
and proved it all immediately.

Weiss hat Newton gemacht aus allen Farben. Gar manches
Hat er euch weis gemacht, das ihr ein Säkulum glaubt.
*
Spaltet immer das Licht! Wie öfters strebt ihr zu trennen,
Was euch allein zum Trutz Eins und ein Einziges bleibt.
*
Es lehrt ein grosser Physikus mit seinen Schulverwandten: 'Nil
luce obscurius' — Ja wohl! für Obskuranten.
*
'Newton hat sich geirrt?' — Ja, doppelt und dreifach. —'Und
wie denn?' Lange steht es gedruckt, aber es liest es kein
Mensch!
*
Schon ein Irrlicht sah ich verschwinden, dich Phlogeston!
Balde, O Newtonisch Gespenst, folgst du dem Brüderchen
nach.
*
Leidlich hat Newton gesehn und falsch geschlossen;
am Ende Blieb er ein Brit: verstockt schloss er, bewies er so
fort.

* (Brücke's comments on Newton's color names, 1866)

An example of a more positive contribution by Goethe about colors, is found in the aesthetic experience described in the following paragraph:

'In traveling over the Harz in winter, I happened to descend from the Brocken toward evening; the wide slopes extending above and below me, the heath, every isolated tree and projecting rock, and all masses of both, were covered with snow or hoarfrost. The sun was sinking towards the Oder pond'.

'During the day, owing to the yellowish hue of the snow, soft violet shadows were to be seen; these were now bright blue because a stronger yellow was reflected from the lighter areas of snow.'

'But as the sun was at last about to set, and its rays were greatly weakened by the dense evening mist, the whole world around me was plunged into a most beautiful purple-red color. The shadow colors changed to green, the lightness and clarity of sea green, with the beauty of the green of the emerald. The spectacle became still more enthralling, one felt transported into a fairyland. every object had clothed itself in these two vivid and so beautifully harmonizing colors, till at last when the sun finally the magnificent spectacle was lost in the gray twilight, and then emerged, slowly, into a clear moon- and starlit night.'

7 The Circle as a Basic Form

Influences from Newton's color ring

Ordering aspects

Scientific aspects

Newton, with his color ring, provided a very valuable contribution, in two respects: first, the basis for further scientific color research with the divided sunlight spectrum (#90 above the black line with the arrow), second, a basis for systematic color ordering in a closed ring of color hues (#13). The relationships between the adjoining colors, the related color hues, as well as the relationship of color hues opposite to each other in the circle, the visually contrasting colors, could be determined.

We can expand certain color areas and contract other color areas in a closed color ring. This results in a displacement of the color pairs which lie opposite to each other. The visually contrasting colors will always, however, be determined by the laws of our color perception.

The circle as a basic form

The straight line (#7 & 8) which we used in linear color ordering from light to dark, from white to black, is followed by the closed color ring as basic form, a circle.

Comparison of various color ordering systems

We will use the circle as a basic form into which we can place a great number of differently shaped color ordering schemes; we can then easily compare the place of the basic colors in the color hue circles. This makes it easier to understand the essential differences between the successive color diagrams.

Almost all symmetrical formations can be placed into a circle, an equilateral triangle, a square, or five-, six-, seven-, eight- and other multi-cornered shapes, (#15-22). The star shapes which are formed by connecting diagonals, and other lines between two colors, can all be positioned within a circle. A further advantage of this circle form is that all full colors are equidistant from the middle point, which is a colorless neutral not influenced by any given color.

All 'full' colors equidistant from the neutral point

The place of the basic hues on the schematic color circle

Color hues at the intersection of lines

There are diagrams in which the basic color hues are indicated by a univocal point such as the intersection of a diagonal, or diameter (#24). Such a point can easily be indicated upon the uniform circle circumference.

Color hue in the middle of a sector arc

There are also diagrams in which the basic color hues are not indicated by a univocal point, but by a color surface, for example a sector, (#33). In this case, the unequivocal color point is placed in the center of a sector arc and this point is designated as the place for a given color hue upon the uniform circle circumference. #15-22 show us how we can use one continuous closed ring of colors in which to place the many differently shaped and composed color ordering systems.

8 Main Divisions of the Basic Colors

The four main divisions of the basic colors of a uniform color circle are listed here:

Four main divisions

1 Newton, the colors of the daylight spectrum, #13.
2 The unique or primal colors, later the basic colors of the opponent-theory, as +/-values by signal-transmission in the color perception process (#23 & 119).
3 The theory derived from very arbitrary, as far as the eye is concerned, paint mixing properties (#30).
4 The main division of the basic colors in agreement with the laws of our color perception possibilities (#37).

These arrangements upon the circle circumference make it possible to compare the essential differences of the individual ordering principles with each other.

Green as a fixed point is always on top

We chose one fixed point to be the same in each color ordering arrangement, the color green at the top in the middle. Green is the color which we perceive when the eye receives light mainly from the middle wavelength area of the daylight spectrum. The color green is top middle, we will place the blues in the left circle half and the yellows in the right circle half.

Some diagrams as mirror images

When in some existing diagrams blue is on the right and the yellow on the left, when green is at the top, we must think of these diagrams as mirror images. The color names of the dots, which will be used to illustrate the diagrams in the 'gray' section of this book, are listed under 'Intended colors' #2.

The names of color researchers, with their own color diagrams within one of the four main divisions, are placed next to each circle arrangement, which has been either mirrored or not for ease of comparison.* There may be differences in the color names used by the given color researchers as compared to the uniform list which we use. When one becomes acquainted with the development of the various diagrams and their descriptions, it is usually clear which colors the color researchers intended.

Correct color images despite confusing color names

Some outdated color ordering principles seem to crop up again and again, sometimes even more than one and a half centuries after their introduction; we can see this clearly reflected in the dates placed by the

* (see the color circles in #19, 23, 30 & 37)

diverse diagrams. Sometimes, we can see, a significant ordering principle was lost, only to be rediscovered after a couple of centuries, and to regain a place in a higher development stage of the color ordering process.

Finally, development leads us to the schematic three-dimensional color-perception-space by Gerritsen, 1975. All colors are schematically organized according to the laws of color perception as to: color hue (the color of a color), the color lightness (how light or dark is a color) and the color saturation (from full color to color(hue)lessness.

9 The Unique Primeval Colors in a Color Ring

Emotional contrast values

The primal colors blue, green, yellow and red (plus the achromatic colors black and white), could be divided, as to the contrast values of the 'primal-instinct' emotion, into pairs which we could place opposite to each other (#23). Blue is opposite yellow, red is opposite green, and later, as the achromats are added to form a three-dimensional structure, black is opposite white.

Blue/yellow contrast

Yellow   Emotional contrast color: might of Chronos, the time, sun, moon and stars; the transitory and time-related.

Blue mighty infinity of the universe and the heavens, meditation and eternity.

Red/green contrast

Red   Emotional contrast color: furious power of fire, blood, war violence, turmoil, movement and ego.

Green force of sprouting seed, power of peace and prosperity, serene rest.

Black/white contrast

Black   Emotional contrast color: power of darkness, death, decay, mourning, destruction.

White power of the light of the world, unspoiled, purity, light of the spirit.

See also Chapter 2: 'The colors as primeval powers' and the color description of these unique colors attributed to Leonardo da Vinci, 1452-1519 (still organized according to the linear principle).

The unique colors in a diagram

Waller's color diagram, 1686, incorporates the contrast color arrangement of the unique color pairs following Newton's color ring. Blue/yellow and red/green lie opposite to each other (#24). The mixed color pairs (mixed paint colors) bluegreen/orange and green-yellow/violet lie between these unique contrast pairs.

Blue and yellow are complementary

Blue and yellow color-perceptions together result in a neutral perception. Red and green color-perceptions together result in a yellow perception.

Red and green are not complementary

The center of the diagram represents neutral — and also yellow perception; this is contradictory. Complementary color pairs are not situated opposite each other, except for the complementary color pair blue-yellow, in this color ordering. The colors cyan and magenta are missing, as are the colors derived from them. This poses a limitation to the available color palette.

The colors cyan and magenta are missing

Hering's opponent theory

This arrangement is reminiscent of Hering's opponent theory, 1834-1918 (#26). He presumes an inversive process in color perception, black-white, blue-yellow and red-green. His opponent theory is not correct as to this formulation. Later, in another context, this opponent theory appears as the +/-values by signal-transmission in the color process. (#119-122).

We find the opponent classification in, among others, the following: Waller—1686 (#24), Höfler—1883 (#25), Titchener—1887 (#70), Ebbinghaus—1902 (#62), Boring—1929 (#66), Johansson—1939 (#74), Plochere—1948 (#27), Hesselgren—1953 (#75), Hård, Natural Color System (NCS)—1968 (#28) and the CIELAB and CIELUV orthogonal system—1976 (#29): yellow/blue, green/red, white/black.

10 Mixing Properties of Paint Colors

Variations of the color circle form which was introduced by Newton

The color ring of the spectrum is irregular

The color ring, such as the one Newton derived from the daylight spectrum, presents a rather irregular impression:

1 The individual color surfaces are of mutually different sizes (the fanned-out daylight spectrum).
2 The color steps between the red and the blue-violet are noticeably great.
3 Two light colors, yellow and cyan, seem to be unduly accentuated.
4 The basic color magenta is missing.

The search for schematic ordering, with the mixing properties of paint colors, has led to misleading conclusions.

Regularity is sought with the mixing qualities of paint color.

The artist looks for regularity and laws of harmony with his paint mixing colors. He still does not know the possibilities of our color perception. He looks for regularity and ordering in the color ring with his knowledge and the means at his disposal. The color surfaces are all given the same size. He looks for visual color steps of equal size. A color ordering from one light color to one dark color is still sought, although black and white are not included in this color ring.

Two color series from light to dark

One color ring is created with two color series from light to dark. These are all placed, however, on one and the same lightness level if compared to the neutral lightness axis between white and black (#31 & 58). One series goes from yellow through orange, red, carmine, red-violet to dark-violet; the other from yellow through yellow-green, green, blue-green, blue, blue-violet to dark violet (#30 & 137).

Newton's light perception color cyan, which appears in the daylight spectrum, disappears. A blue-green appears, between green and blue, it is darker than the green and lighter than the blue. It was not yet known that ultra-marine blue and red can each be compounded of two pigments. Ultra-marine blue, yellow and red were accepted as the three undividable primary colors. The pigments used today for three-color printing, color film and color photos, namely cyan and magenta, had not yet been discovered.

Magenta and cyan were missing

The colors, made by mixing two of the presumed primary colors ultra-marine blue, yellow and red, were called secondary colors. The colors magenta and cyan could not be produced. This is how the presumed (but false) basic secondary colors orange (of yellow and red), green (of yellow and blue) and violet (of carmine red and ultra-marine blue) came to be promoted.

Continued on page 49

The diagram schemes can be 'colored' with color samples

After the introduction given in Chapter 1, we will now provide further instructions on illustrating the diagrams with color samples.

The numbering of the color samples

The numbers will indicate the right place, in the illustrations, for each colored spot. The color spots are numbered consecutively. If all yellow spots had been given the same number, and all blues, reds greens, etc, another number -one per color- a sham relationship between a number and a given color could be presumed; this we must avoid. Hickethier's system does use this type of color coding. One could utilize Hickethier's color number code, as a comparison, and write them next to the colors used here; but in some cases one will find no Hickethier number for a given color spot.

There are l90 color spots on the color chart (page 79), enough for all the illustrations. Take care when cutting the dots. Perhaps it is a good idea to practice cutting round samples out of old magazines, advertising material etc. If one of the 190 color spots is spoiled, it can be replaced by a matching sample from one of the many advertising folders e.d. that turn up in our mailboxes every day.

The paper punch

A 10 mm punch can be bought in a hardware store or hobby shop. Be sure to choose the right diameter, 10 mm, usually a '10' is stamped on the shaft. The blade should be round, straight and without nicks. Try the punch out while still in the shop; take along a sample of colored paper and a piece of gray cardboard to be used as a punching base. Make sure that the diameter is 10 mm and that the cut samples are neat and round.

The punching base

We must use a piece of thick gray cardboard, such as from a mailing box, as a base. Yellow strawboard cannot be used as it contains grains of sand which will quickly dull the blade. The punch must be moved, in relation to the cardboard, as often as necessary. Place the cardboard on a strong hardwood plank (e.g. the breadboard). Place the plank on a protective layer on a solid table, perhaps just above the table leg for extra support.

The hammer

An ordinary hammer can be used for this work. The top of the punch may show burrs after intensive use; these are easy to file away. If a hardwood or plastic hammer is used there will be no burrs.

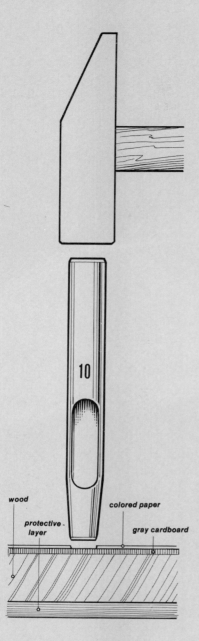

#1 Punching the color samples with a punch

Practice first

First it is best to practice punching on colored waste paper. The punch must be held absolutely vertical. Then one hits the head of the punch with a short, firm stroke. The paper circles often remain in the punch. They can be pushed out easily by means of a cotton swab or similar object inserted into the opening of the punch.

The color sample page (pp. 79-80)

Place a thin sheet of cardboard, such as the back of a pad of writing paper, between pages 80 and 8l. Cut the color page out of the book with a sharp knife (stanley knife), cutting along a metal ruler. Use the dotted line at the left edge of the page as cutting line so as not to damage the last row of samples.

Punching and glueing

One can see on each diagram which color spot(s) should be punched. Set the punch so that it does not touch the black border; then with one firm stroke punch the color dot out and glue it in place immediately. Make sure that the number on the back of the spot agrees with the number printed on the diagram. Use a good paper glue; sticks of glue are not practical for this job because this paste dries out after some time and the samples will come loose. It is best to use a thin glue from a tube; transfer a little glue with the fingertip or match stick to the dot and glue in place immediately. Thick wallpaper paste can also be used. It will leave less spots.

Experimenting with colors

First you must illustrate this book; and then, now that you own a punch, you can also experiment with punched-out color samples of any available colored paper, cellophane, plastic, leather, textiles etc. This will encourage you to discover new color nuances and combinations. Try to find the place of any given color in the color space. Look at pages 75, 76 and 77, which colors match the given samples? Which are redder, yellower, lighter, darker, more or less colorful? It is even better to find the place for an arbitrary color in our self-constructed color-perception-space model Gerritsen-l975, which has been 'colored' with a schematic range of colors (see page 88).

#2 The color names of the basic colors

Here are different names one will see used for some main colors + the neutrals or achromats: white, the grays and black:

colors intended		many names often confusing
violet	86	violet, purple
blue	33	blue, ultra-marine, violet-blue, blue-violet, violet
cyan	60	cyan, cyan-blue, blue, turquoise, green-blue, light blue.
green	5	green, blue-green, yellow-green
yellow	112	usually univocally called yellow
orange	167	usually univocally called orange
red	139	red, orange-red, orange, carmine
magenta	181	magenta, magenta-red, purple, hot pink, pink, red
white	1	usual univocal name for the lightest achromat
grays	17	grays, neutrals, achromats, gradations between black and white
black	4	usually univocally called black, opposite of white

#3 **The unique colors according to Leonardo da Vinci (1452-1519)**

A negative selection is used in order to help us understand exactly which colors we mean and what their names are. Leonardo da Vinci was the originator of this method of describing colors. A given color can be described with no more than six unique color concepts.

White (15)   White contains nothing of black, nothing of blue, nothing of red, nothing of green, nothing of yellow.

Yellow (113)   Yellow contains nothing of black, nothing of blue, nothing of red, nothing of green, nothing of white.

Green (6)   Green contains nothing of black, nothing of blue, nothing of red, nothing of yellow, nothing of white.

Red (140)   Red contains nothing of black, nothing of blue, nothing of green, nothing of yellow, nothing of white.

Blue (34)   Blue contains nothing of black, nothing of red, nothing of green, nothing of yellow, nothing of white.

Black (110)   Black contains nothing of blue, nothing of red, nothing of green, nothing of yellow, nothing of white.

THE THREE DIMENSIONS OF A COLOR COLOR HUE, LIGHTNESS, SATURATION

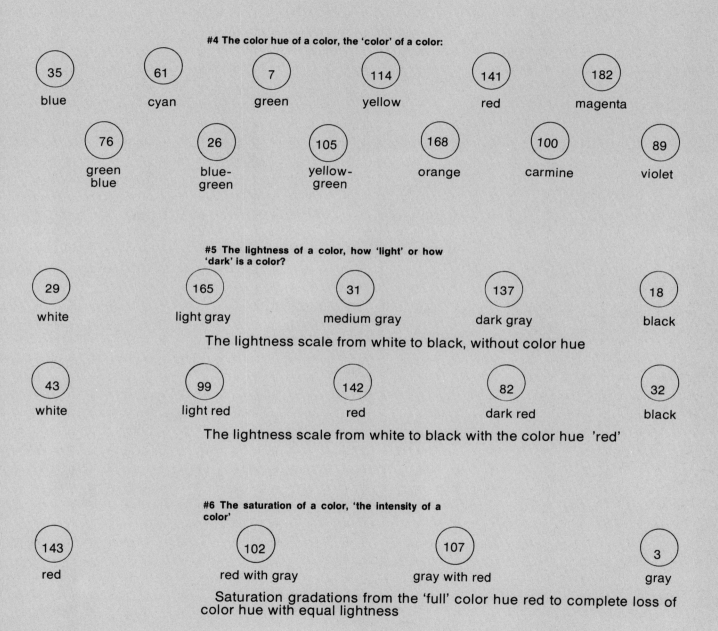

#4 The color hue of a color, the 'color' of a color:

35	61	7	114	141	182
blue	cyan	green	yellow	red	magenta

76	26	105	168	100	89
green-blue	blue-green	yellow-green	orange	carmine	violet

#5 The lightness of a color, how 'light' or how 'dark' is a color?

29	165	31	137	18
white	light gray	medium gray	dark gray	black

The lightness scale from white to black, without color hue

43	99	142	82	32
white	light red	red	dark red	black

The lightness scale from white to black with the color hue 'red'

#6 The saturation of a color, 'the intensity of a color'

143	102	107	3
red	red with gray	gray with red	gray

Saturation gradations from the 'full' color hue red to complete loss of color hue with equal lightness

#7 Linear color ordering from light to dark, between white and black

'The mixed color between white and black is red (ancient Greeks): between the 'white' of the day and the 'black' of the night is the red of the rising and setting sun.'

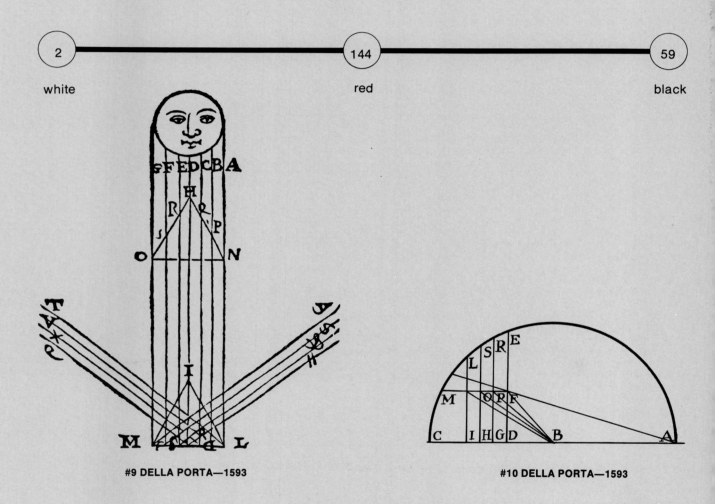

| 2 | 144 | 59 |
| white | red | black |

#9 DELLA PORTA—1593

#10 DELLA PORTA—1593

Colors originate from the struggle between darkness and light

This linear color ordering from light to dark, between white and black, represents the ancient Greek conception that the colors originate from the struggle between light and darkness. The mixed color between white and black was presumed to be red; between the light of day and the black of night is the red of the rising and setting sun. The other unique colors were therefore ordered as to their 'own' light/dark values. The light yellow lies between red and white, the darker blue between red and black.

#8 The 'unique' colors from light to dark between white and black, ordered according to their own light/dark values

Aristotle's false theory, that a mixture of blue and yellow light is green light, is prolonged here. We now know that the mixed color of yellow and blue light is white, and not green.

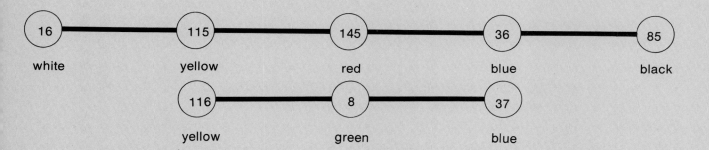

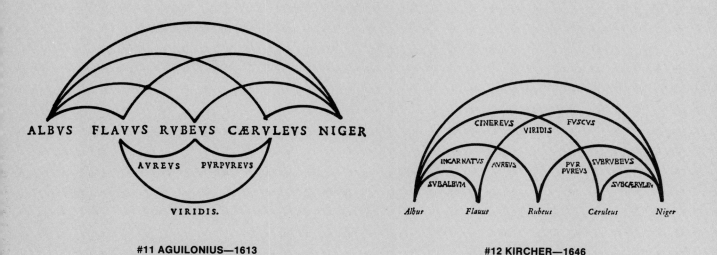

#11 AGUILONIUS—1613

#12 KIRCHER—1646

The own light/dark value of green is between the lighter yellow and the darker blue. Aristotle's incorrect theory was that green was the mixed color of yellow and blue light. Today we know that mixing yellow and blue light results in white, not green. As yet, no thought was given to color contrast values in this development phase, but there was the positive fact of color ordering as to the colors' own intrinsic lightness values.

The color-contrast values can not be seen in this color order.

**The colors of the daylight spectrum
The color-hue ring as an ordering principle**

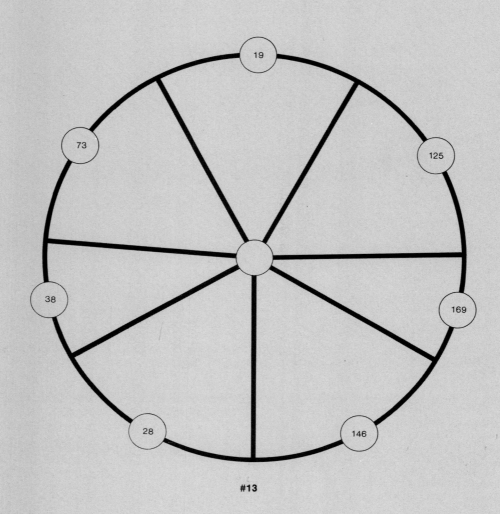

#13

Newton divides sunlight through a prism

Sir Isaac Newton broke with the existing color ordering from white to black. He divided sunlight, with the aid of a glass prism, into the colors of the daylight spectrum; the colors were fanned out into successive wavelengths as was determined two centuries later by Hertz. The colors hues which we perceive are: violet-blue, ultra-marine blue, cyan-blue, green, yellow, orange and red. (Comments on Newton's color names by Brcke, 1866.) Newton gave these colors the following names: violet, indigo, blue, green, yellow, orange and red. He placed these colors into a closed circular color ring. White was in the center of this circle formed by the color ring.

Sir Isaac Newton 1642-1725

Influences from Newton's color ring

Newton in two ways provided a valuable contribution with his color circle:

Scientific aspect

He laid the foundation for further scientific color research with the division of the daylight spectrum as a research tool.

Ordering aspect

He also gave us a basic scheme for systematic color ordering in a closed ring of colors. The relationship between colors next to each other, the related colors, and also the relationships between colors which lie opposite to each other in the circle, the visually contrasting colors, can be determined. It is possible to expand some color areas and contract others within a closed ring of colors. A displacement of the color pairs, which lie opposite to each other, takes place. The visual contrast colors will always by governed by the laws of our color perception.

Ordering as to light/dark values of colors was lost

Black and white are no longer included in the new circular color sequence. Color ordering as to light/dark values, the lightness values of the color hues in relationship to black and white, which was used for centuries, is lost in Newton's color circle.

The lack of light/dark relationship between the color hues is a significant loss for the (color)artist. This is because the light/dark contrast has greater visual contrast effect than a color contrast of equal lightness. It is, therefore, important to know how light or how dark a color hue is. This loss, which occurred in Newton's color ordering, was hardly noticed and certainly never analyzed.

A period of more than 2200 years came to an end, an era in which uncounted color scientists ordered colors on a straight line from the lightest to the darkest color, starting with white and ending with black.

It was not until the Twentieth Century that the lightness of the basic colors was again included in the classification of colors, also according to their own intrinsic lightness in relation to the lightness gradations from black to white.

Green at the top as a fixed point of comparison

The four basic-color-hue circles with colored dots (#13, 23, 30 & 37) have been so oriented that the color green is always centered at the top; this is as in Maxwell's triangular scheme and the CIE color triangle. Thus we gain a fixed point for the mutual comparison of the various arrangements.

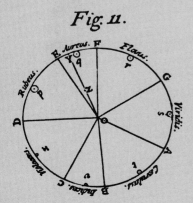

#14 Newton's color circle from his Latin publication Optice, London, 1706.

The circle as basic form enables on to make mutual comparisons between the different color-ordering principles in diagrams of very different shapes.

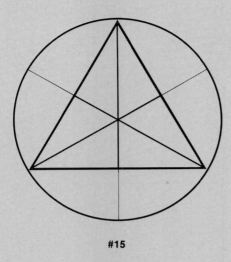
#15

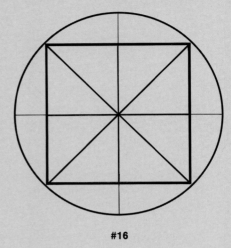
#16

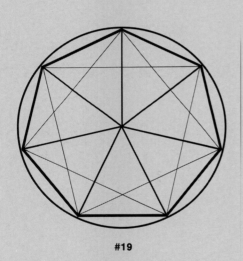
#19

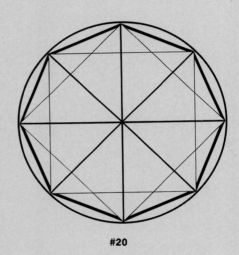
#20

The circle as basic form

We see how color orderings made up of one closed ring, but with very different shapes and totally different connecting lines between the various colors mutually, can be contained within our chosen circle form: #15 through #22 00. One advantage of the circle form is that all full colors are equidistant from the center, the colorless neutral which has no inclination towards any color hue. There are diagrams in which the basic color hues are indicated by a diagonal or diameter (#24). A univocal point can then be indicated on a uniform circle circumference. We will place the univocal color point in the middle of a sector arc of the uniform circle circumference, (#33), in a diagram with color sectors. In the four main divisions of basic

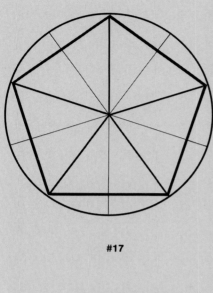

#17

#18

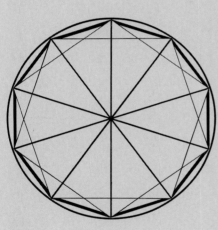

#21

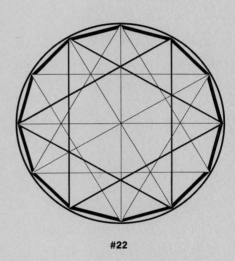

#22

colors on the uniform color circle: the spectrum colors, the unique colors, the paint mixing colors and the basic perception colors, we always place the color green at the center top in order to have a mutual fixed point (#13, 23, 30 & 37). We will place the blues in the left circle half and the yellows in the right circle half with the color green at the top in the middle. If a given diagram has blue on the right right and yellow at left once the green has been placed at the top, we must mirror-image this diagram. The color names which we give to the samples used to illustrate the diagrams, are found under 'intended colors', (#2).

The unique primal colors in the color

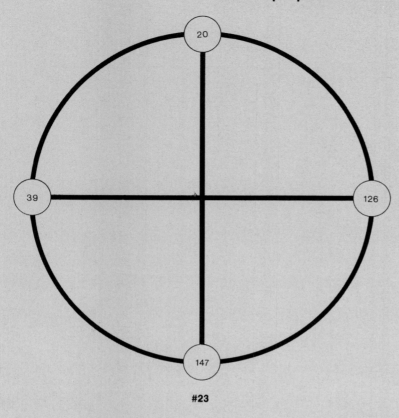

#23

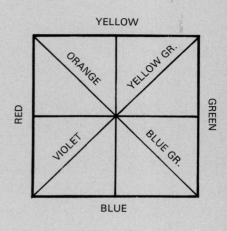

#24 WALLER—1686

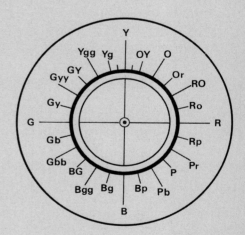

#27 PLOCHERE—1948

The opponent-theory as a basis for ordering color hues

We see the unique colors as contrast pairs blue/yellow and green/red, with the paint mixing color pairs violet/yellow-green and orange/blue-green (which lie between these unique color pairs) along with the neutral center black/white, for the first time with Waller (#24). Blue-yellow excepted, the complementary color pairs do not lie opposite to each other in this color ordering. The colors magenta and cyan are missing here, along with the colors derived from them. This ordering can be regarded as an early precursor of Hering's opponent-theory of 1911; this the reversal process: black-white, blue-yellow and green-red is supposed to occur in color vision. A scheme (In #119) of +/-opponent values of signal transfer in (color) vision is shown, with the complementary basic-perception-colors opposite to each other in the color circle.

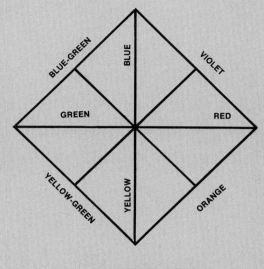

#25 HOFLER—1883

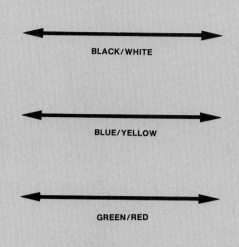

#26 HERING—1911

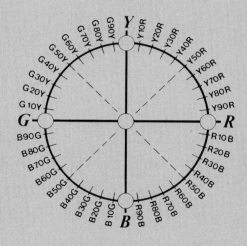

#28 NATURAL COLOR SYSTEM NCS—1968

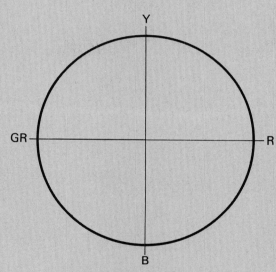

#29 CIELUV—CIELAB—1976

Mixing characteristics of paint colors as a basis for color ordering

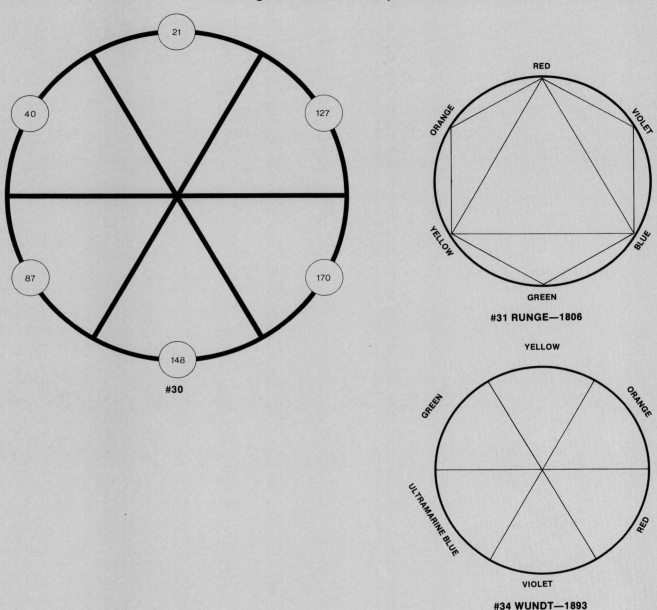

#30

#31 RUNGE—1806

#34 WUNDT—1893

A misleading basis for visual color contrasts

The artist uses the colors of the daylight spectrum. He divides a circle into six color sectors of equal size. He thinks that the primary colors are ultra-marine blue, yellow and red; he names their mixed colors secondary colors: the green, the orange and the violet; he then concludes that the color pairs blue-orange, yellow-violet and red-green are visually complementary colors, but this is not true. The light spectral color cyan has disappeared, and perception color magenta is also missing. Complementary color pairs do not lie opposite each other in this color circle.

Philipp Otto Runge's color circle, taken from his letter to Goethe of July 3, 1806. The color circle arrangement by Vincent Van Gogh (#33) is taken from his letter (No.116) to his brother Theo (circa 1878), in which he expresses the views of his contemporaries.

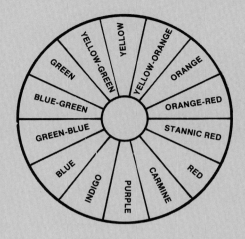

#32 HERSCHEL—1817

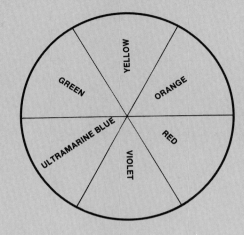

#33 VAN GOGH—1878

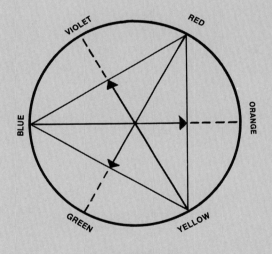

#35 KLEE—1924

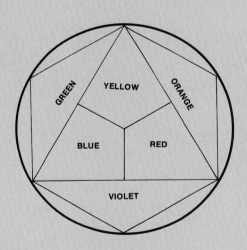

#36 ITTEN—1961

Laws of our color perception possibilities as a basis for color ordering and diagrams which show this agreement

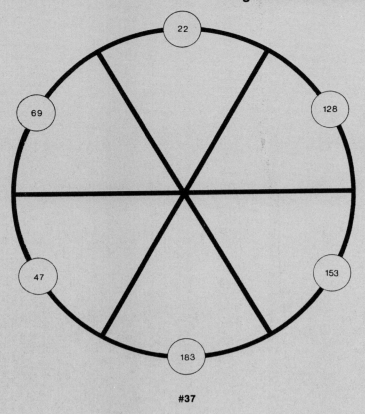

#37

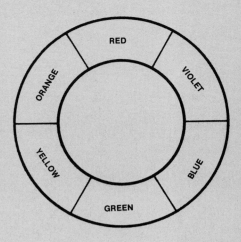

#38 GOETHE—1793

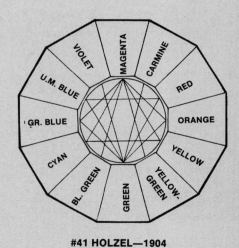

#41 HOLZEL—1904

A reliable guide to determining and influencing visual color contrasts

Colors originate when we see them and are therefore bound to the laws of our color perception. We can learn to know these laws through research into the relationship between light of a given wavelength area and the color perceived. This is also true for paint colors, the reflection of a color surface is light of a certain wavelength which we perceive as the color of that reflected surface. This is how we arrive at the hypothetical eye 'primaries', ultra-marine blue, green and red; and the hypothetical eye 'secondaries', cyan, yellow and magenta. The color ordering originating from these colors does offer the visual complementary colors opposite each other. The colors magenta and cyan are utilized as basic perception colors.

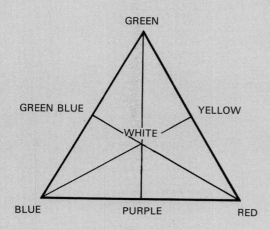

#39 MAXWELL—1857

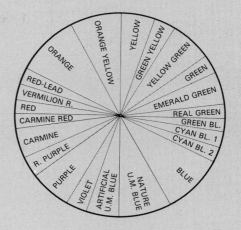

#40 ROOD—1879

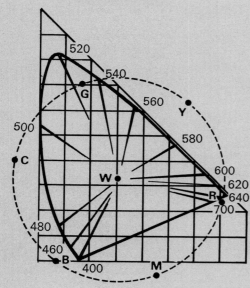

#42 CIE COLOR TRIANGLE—1931

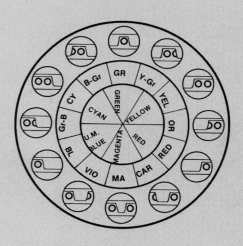

#43 GERRITSEN 12 SECTORS—1975

From full color hue to white, from full color hue to black. In this color circle the complementary color pairs lie opposite to each other

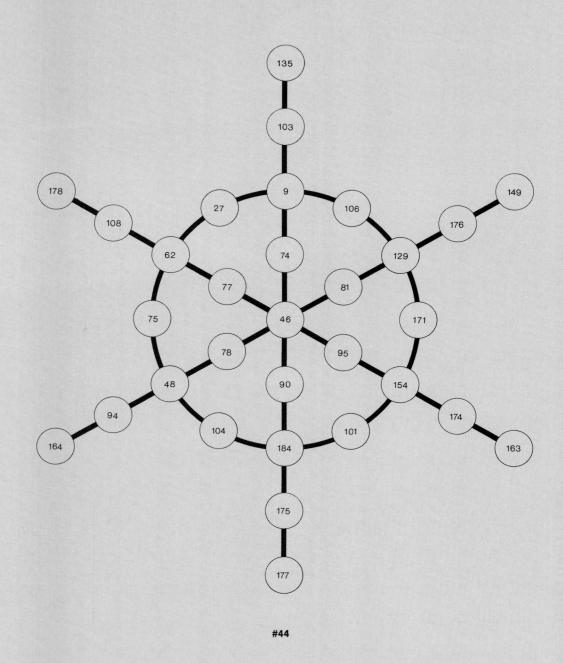

#44

There is no possibility of mixing all colors with all grays between white and black in this color ordering.

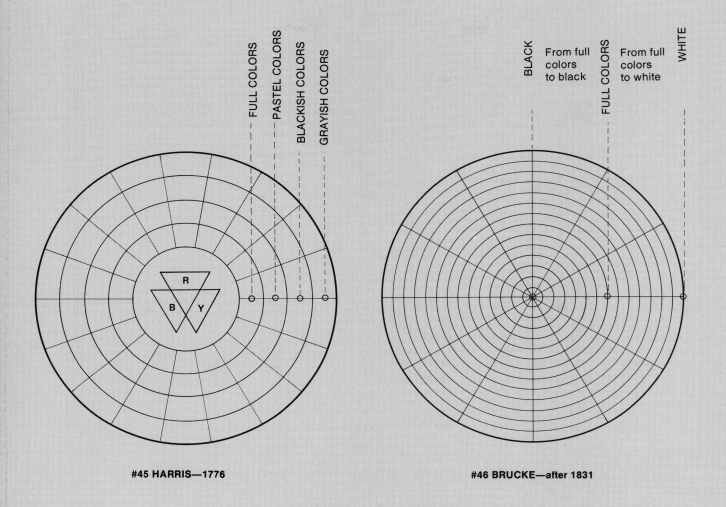

#45 HARRIS—1776   #46 BRUCKE—after 1831

Moses Harris's color circles

Moses Harris made various color circles, in 1766, with three color tone plates in diverse gradations of red, yellow and blue. He made the eighteen-part color circle above, with color rings and a total of 72 colors, in 1776. The inside ring contains the pure color hues. The next ring shows the pastel tints, color mixed with white. Then comes the ring of dark colors, color mixed with black. The outer ring contains the grayed colors, color mixed with gray.

Brücke's color circle

Brücke made a color ring of six pairs of complementary color hues. There are seven concentric color rings inside the 'pure color ring', each successively darker to the black point in the middle; there are seven concentric rings outside it, ending in the outer ring which is white.

These color circles, by Harris and Brücke, do not allow for the possibility of mixing all colors with all grays between white and black. The color diagram must be three-dimensional to make this possible.

All color mixing possibilities between the full color hue and white, all grays and black. Color ordering becomes three-dimensional.

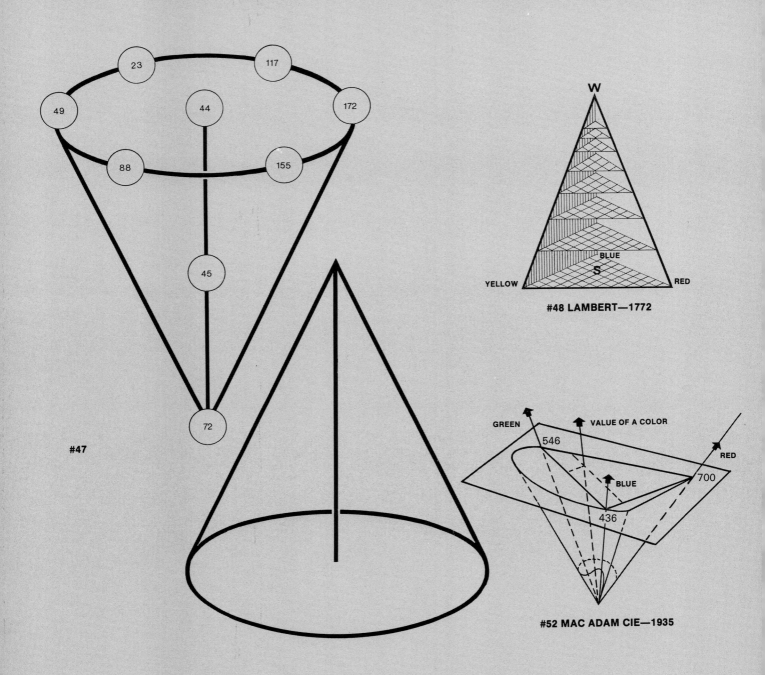

#47

#48 LAMBERT—1772

#52 MAC ADAM CIE—1935

The distance between the full color hue and white and the distance between full color hue and black is not equal. The mutual color steps are therefore not of equal size.

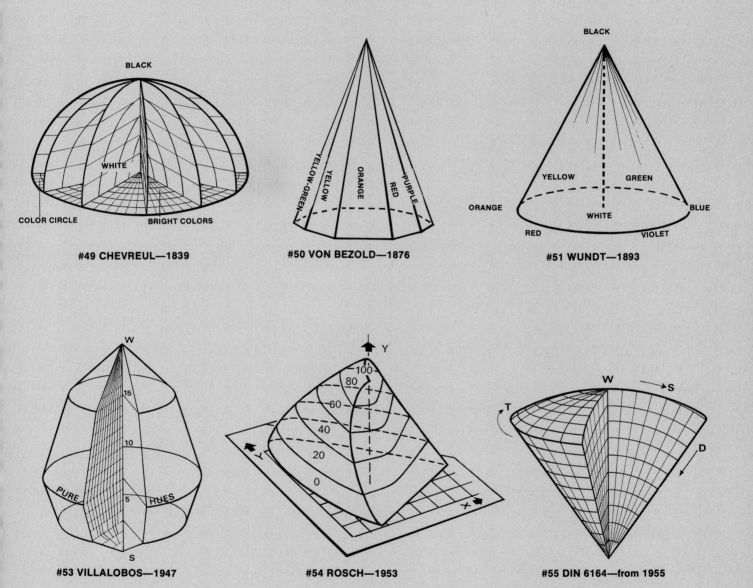

All color hues of the color wheel have the same distance from white and black

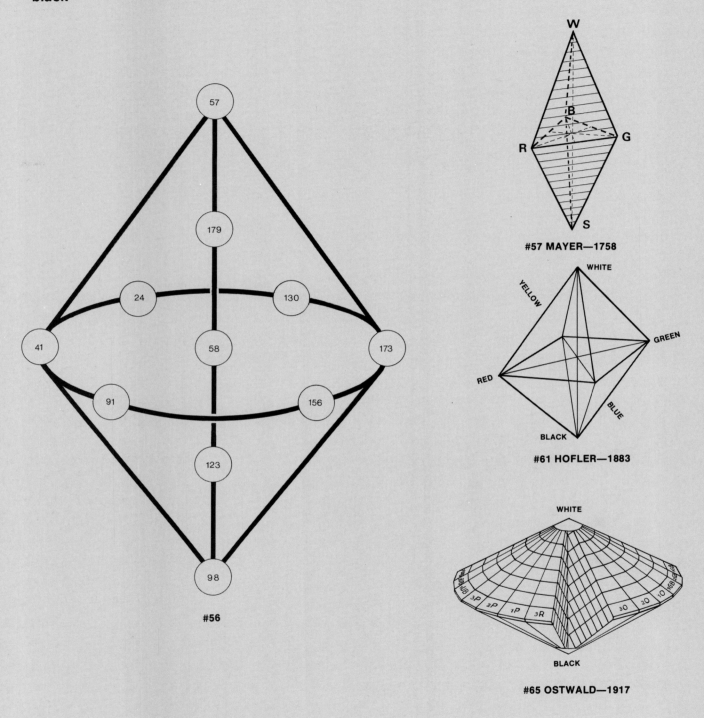

#56

#57 MAYER—1758

#61 HOFLER—1883

#65 OSTWALD—1917

All full colors, in this color ordering, are not placed correctly as to their light-dark value relative to white and black

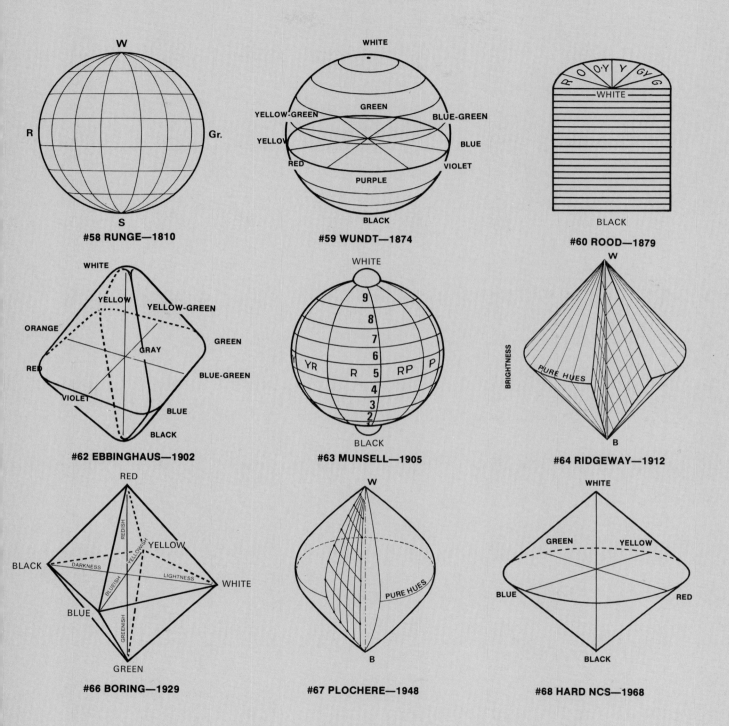

The light color hue yellow lies closer to white and further from black. The dark color hue blue is nearer to black and further from white.

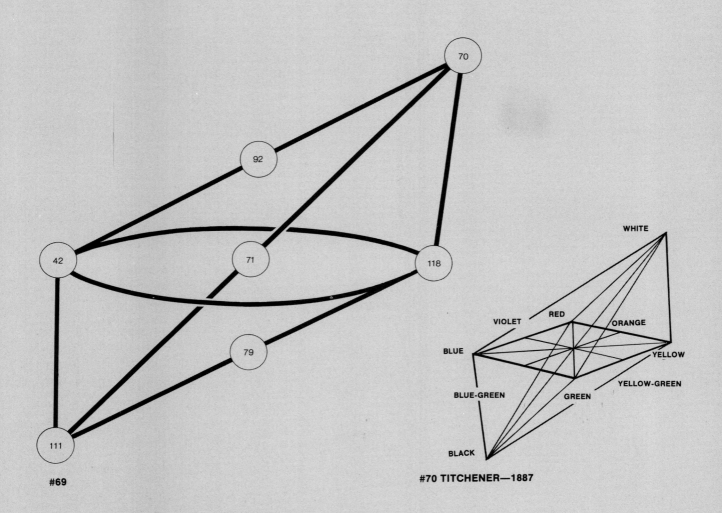

#69

#70 TITCHENER—1887

Titchener, at the end of the 19th century, was the first, after the ancient Greeks, to make an attempt to organize the color hues according to their own lightness/darkness value, relative to black and white. He placed the lightness axis at an angle through the center of the horizontal color hue ring. Thus, the lighter color hue yellow is closer to the white point and farther away from black point. The darker color hue blue, on the other hand, moves closer to the black point and farther away from white point.

The ordering of basic color hues according to their own light/dark values, relative to white and black, is recovered once more, after being lost since Newton.

Albert H. Munsell, 1858-1918

From one light color to one dark color

The American painter Albert H. Munsell created a spherical color space in 1905. The color hue ring is still horizontal here, so that the light yellow and the dark blue lie on one lightness level. He created, in 1915, his brilliant color diagram from which the now world-famous Munsell color system developed. He leaves the neutral-axis, with light/dark values from black to white, vertical and places the color ring at an angle. The lightest color hue, yellow, is above and the darker blue-violet is below. This color circle, in 1915, was mainly ordered from the light color yellow to the darker color blue-violet in two directions: one series of colors via red from lighter to darker color hue: yellow, orange, vermilion, carmine, red-violet, blue-violet; and the other series of colors via blue from lighter to darker color hue: yellow, yellow-green, green, green-blue, ultra-marine blue, blue-violet. The slanting ring of colors actually consists of the saturated full color hues which lie at the outer ends of the color arms. These color arms point horizontally away from the lightness axis in the direction of a color hue. The lightness of the color hue remains constant; only the saturation, the intensity, of the color increases as it is further removed from the colorless neutral-axis. We see, in #123, a vertical cross-section of a color space through the complementary color pair yellow/ultra-marine blue. The length of the various color arms is determined by the different lightness levels as to the intrinsic lightness of the color hue (#72).

The Munsell color system, in 1915, did not sufficiently take Maxwell's basic color arrangement into account with the choice and direction of the color hue samples.

The Munsell color system of today provides room for a much greater assortment of color saturations. The system is brought up to date, from time to time, in accordance with significant new developments. This is possible because the fundamental design of this color ordering principle by Albert H. Munsell.

Here all color hues are again ordered according to their own light/dark values, relative to the light/dark values of the neutral axis from black to white.

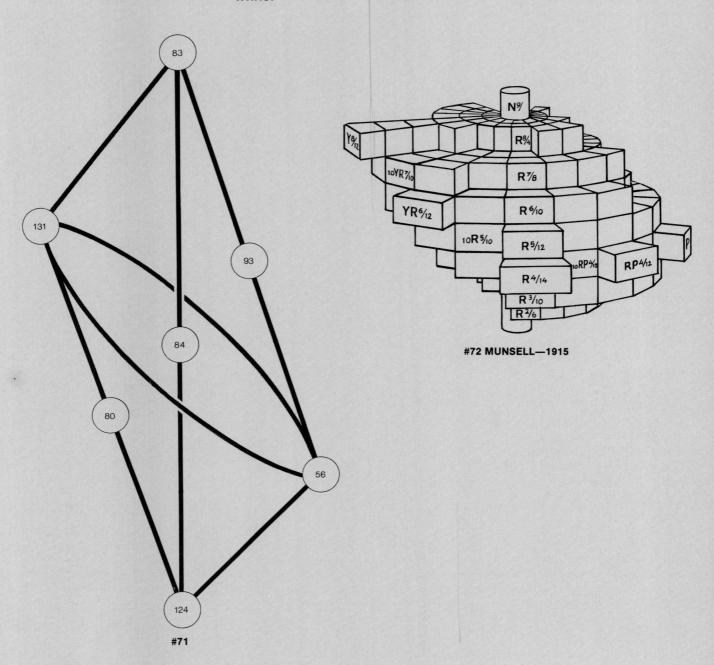

#72 MUNSELL—1915

#71

This color ordering from one light color, yellow, to one darker color violet-blue, does not, however, take the Maxwell color scheme into consideration

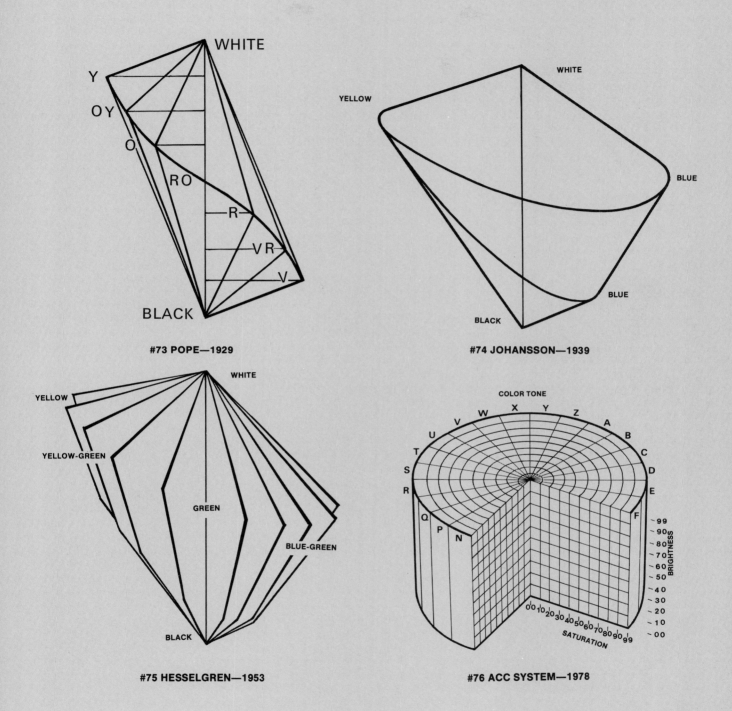

Maxwell's three light and three dark colors in a three-dimensional color scheme

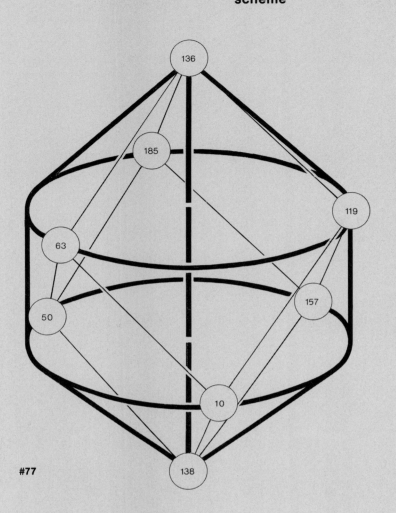

#77

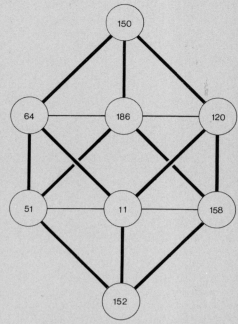

#78 Vertical projection

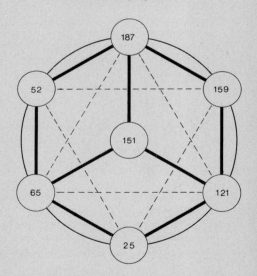

#79 Horizontal projection

Maxwell's three light and three dark colors, alternating in sequence, are seen in a three-dimensional color scheme by William Benson, 1868, in his cube with the additive mixing colors: yellow, cyan (sea green) and magenta (purple); and ultra-marine blue, green and red. Hickethier, 1940, took this cube as the basis for his thousand-color cube for printing inks. He used very confusing color names, he calls magenta: red, red: orange, cyan: blue and ultra-marine blue: violet. Only yellow, white and black were named correctly. The alternating sequence, of three lighter and three darker colors, as is found in Maxwell's and Benson's schemes, is maintained. The three lighter colors, yellow, magenta and cyan, with very different light/dark values, are placed on the same lightness level relative to black and white, which is not correct.

The three lighter colors: yellow, magenta and cyan, with their varying light/dark values, are placed on one and the same lightness level.

The three darker colors: ultra-marine blue, green and red, with their own different lightness levels, are also placed on one and the same lightness level.

Color and lightness are, therefore, disorieted throughout the whole color space of this ordering principle, the cube and the rombo der.

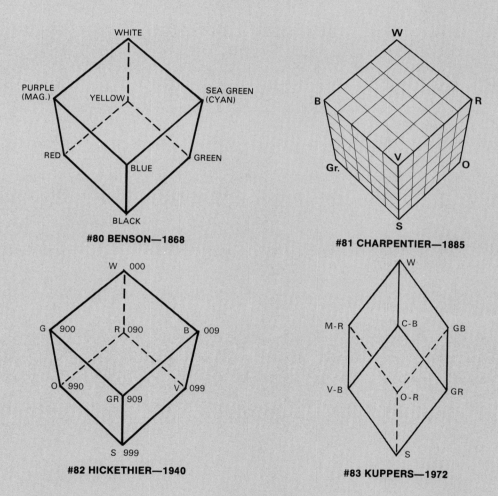

#80 BENSON—1868

#81 CHARPENTIER—1885

#82 HICKETHIER—1940

#83 KUPPERS—1972

The same mistake is made with the three darker colors: ultra-marine blue, green and red; three colors of different light/dark values are placed on the same level of lightness in relationship to black and white.

There is, again, a disorientation of color and lightness through the whole color space of this color ordering principle: in both the cube with its six interfacing squares, and the rhombohedron with its six diamond-shaped surfaces.

The schematic color perception diagram according to the laws of our color perception, Gerritsen 1975

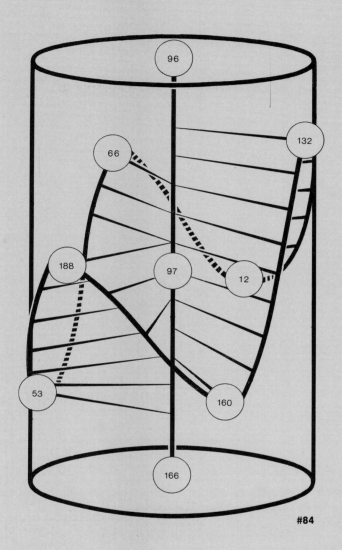

#84

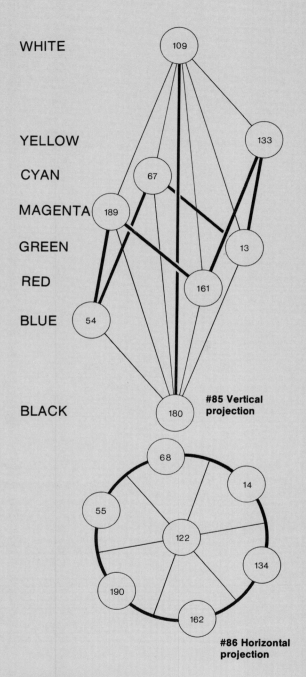

#85 Vertical projection

#86 Horizontal projection

This schematic color-perception-space orders all colors according to the laws of our color perception: color perception and color hue, color perception and degree of lightness, color perception and degree of saturation. We rediscover many old values in this ordering principle:

• The colors are, again, ordered according to their own lightness between black and white, as with the Ancient Greeks.

- The 'mixed color' red is again placed between black and white as in olden times; we find red on the outer perimeter between black and white.
- Newton's daylight spectrum colors are in their proper places.
- The basic colors, three lighter and three darker ones, alternate with each other, as in Maxwell's color scheme.
- The opponent value schemes of these basic colors are found in #119.
- The irregular zigzag line, on the cylinder wall, follows the intrinsic lightness values of the full color hues and guarantees equal lightness and color gradations.
- All complementary color pairs lie diametrically opposite to each other.

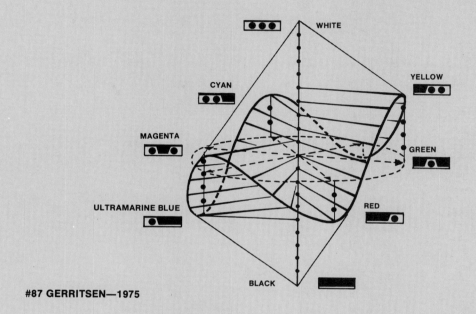

#87 GERRITSEN—1975

A transparent three-dimensional color model of the color-perception-space

A description of how to build a model of the Gerritsen color-perception-space, with schematic color indications, is found in Chapter 25. Read the description carefully. There is a sketch of this transparent color model on page 48, #88.

A drawing at the upper right, #89 A, must be drawn to the actual size of the glass plate upon which the color spots will be pasted; the underside of this drawing shows a dotted line which indicates how deeply the glass plate fits into the wooden base. The half circles indicate the place for the gray sequence from white to black. These gray sequences of half circles will be the same for all the glass plates. #89-B shows the same drawing as #89-A with with all possible little circles or dots; these must be placed in the same horizontal pattern as the half circles.

We can attach all the color samples to their correct places according to this one drawing. Each new glass plate, of another color hue, can be placed onto this same drawing. The colored dots, which we have made ourselves, and which have been colored on both sides according to the description given, can be pasted into their correct places; this according to the uniform pattern for all color hues on the lateral cross-section halves #89 C and #125-126-129 through #134.

Model of the color perception space made from glass plates with color samples pasted onto both sides.

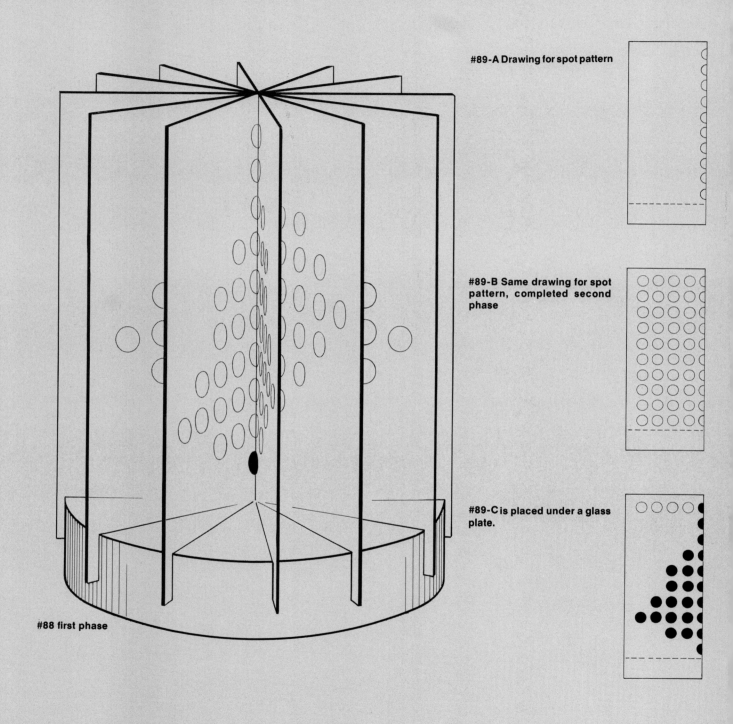

#89-A Drawing for spot pattern

#89-B Same drawing for spot pattern, completed second phase

#89-C is placed under a glass plate.

#88 first phase

Every color spot appears the color samples matt on one side (paint) and glossy can be glued onto the on the other, seen through glass plate in the the glass special pattern for each color hue

48

Continued from page 16

Primary colors chosen arbitrarily with regard to the eye

Complementary color pairs are not opposite each other

This color hue ring shows main colors which, in as far as our color perception possibilities are concerned, have been chosen arbitrarily: ultra-marine blue, green, yellow, orange, red and violet. These basic colors were wrongly chosen; the colors which lie opposite to each other, on the circle constructed from these colors, were called complementary i.e. were presumed to form a neutral together. These colors are: ultra-marine blue/orange, yellow/violet and red/green. It is thought that all color pairs, of colors which lie opposite to each other, represent all three main colors: ultra-marine blue, yellow and red. (Later we shall see that the eye primaries are: ultra-marine blue, green and red (#94, 97, 102, 103 & 104), and the eye secondary colors are: yellow, magenta and cyan (#99, 100 & 101)).

Red and green are not a complementary color pair

When we see a red and a green light spot overlapping each other on a surface, we can observe that a yellow appears and not a neutral (white, gray, black) (#142). Ordinary red and green paints, when mixed, make a brown color and not a neutral gray. Blue and orange light mixed also do not produce a neutral gray, but a light pink. Yellow and violet light colors also mix to a light pink. Yellow and violet paint mixed produces a green or bronze-green.

A deceptive guide

This color ordering, derived from the mixing properties of paint colors, is deceptive. One cannot learn which colors enhance each other visually, or which colors have the greatest visual color contrast. This arrangement of colors also limits the available color palette. Magenta and cyan, and their related colors, are completely missing.

Limitation of the color palette

There were many artists who were completely convinced that this now outdated and misleading theory agreed with the laws of our visual color perception; some of these artists were: Runge-1806 (#31, 58), Herschel-1817 (#32), Vincent Van Gogh-1878 (#33), Wundt-1893 (#34, 51), Klee-1924 (#35), Itten-1961 (#36). This did not prevent them from creating art works of the highest standard despite the limitations of their color theory.

11 Color Perception as a Basis

Colors exist only by being seen; therefore they are bound by color perception possibilities of our visual organ.

Combinations of light wave lengths

Laws of color formation are studied by combining light colors from the fanned out daylight spectrum, along with studies of paint color mixing properties.

The Young-Helmholtz three component theory

Basic colors blue, green and red

Young (1773-1829) showed that all colors could be produced with only three colors of the daylight spectrum. He draws the conclusion that three types of color receptors suffice for the eye; he called them particles. Helmholtz (1821-1894) later expanded Young's theory; just like Young and Maxwell(1831-1879) did, he calculated curves for the basic light mixing colors: ultra-marine blue, green and red. He writes, as did Young, that the three basic colors and their mixing colors are received by three kinds of color-sensitive receptors (cones). This is known as the Young-Helmholtz three-component theory (see also #93).

Maxwell

Maxwell systemizes three spectral main colors into a triangular diagram (#39) with which he could form all colors. The three colors are:
— in the short wave area of the daylight spectrum, light energy which generates the color sensation ultra-marine blue (the unique blue);
— in the middle wave area of the daylight spectrum, light energy which generates the color sensation emerald green (the unique green);
— in the long wave area of the daylight spectrum, light energy which generates the color sensation red (the unique red).

When short wave light (blue) converges with middle wave light (green) the color sensation of cyan appears (#100-D & E, & 142). Cyan is lighter than either of the two colors from which it originates. Cyan is also found in the fanned-out daylight spectrum, where the short- and middle wave spectral sensitivities of the eye overlap (#90).

When we converge middle wave light (green) and long wave light (red) of the daylight spectrum, the color sensation yellow appears (#99-D & E, & 142). Yellow is also lighter than either of the two colors from which it originates. Yellow also occurs in the divided daylight spectrum; this occurs because the middle and long wave spectral sensitivities of the eye sensitive for light energy, partially overlap.

Finally, when short wave light (blue) and long wave light (red) converge the color magenta (#101-D & E, & 142) appears. This color is also lighter than either of the two original colors blue and red. Magenta does not occur in the divided daylight spectrum. If, however, we withdraw the middle wave light (green) from the daylight spectrum, then the color sensation magenta appears: blue perception plus red perception (#101-D & E, & 142).

An infinite number of mixed colors can be realized by varying the intensity of the three basic light colors ultra-marine blue, green and red. Maxwell's color diagram (#39), shows the complementary color pairs which lie diametrically opposite each other. The complementary color pairs for the six basic colors are: ultra-marine blue/yellow, green/magenta and red/cyan.

The results of experiments made by Maxwell are essentially the basis of present-day three-color printing (plus black), of color film and color photos, of color television and the current CIE color measuring system.

The conclusion which we can draw from these color perception experiments is that colors originate only when we see them. Laws governing colors are therefore bound by the color perception possibilities of our visual sense: eyes, brains, consciousness. See also chapter 18, 'The color-perception-space' (Gerritsen 1975). Color ordering systems which show general agreement with color ordering according to the laws of color perception are Goethe-1793 (#38), Maxwell-1857 (#39), Rood-1879 (#40), Hölzel-1904 (#41), CIE Color Triangle-1931 (#42), Gerritsen-1975 (#43, 117, 119 & 124).

Light wave lengths as a color scheme

Basic mixing colors: cyan, yellow, magenta

Complementary color pairs: ultra-marine blue/yellow, green/magenta, red/cyan

12 From Full Color Hue to White—From Full Color Hue to Black

Newton placed his color ring around a white center (#13), which symbolizes all colors (wave lengths of the daylight spectrum) brought together again as white light.

Moses Harris's color circle

Moses Harris made a color circle, in 1776, which consisted of four color rings around a white center (#45). He placed the presumed paint primary colors ultra-marine blue, yellow and red in the circular white center. He placed an eighteen-part ring of full color hues around this center. Just outside this ring is a second color ring of pastel colors. The third ring contains colors mixed with black. The outermost, the fourth color ring, contains the colors mixed with a neutral gray. There are four color rings next to each other: the full colors hues, only one white gradation, only one black gradation and the gray gradations.

Color rings: full color hues, mixed with white, mixed with black and mixed with gray

Brücke's color circle

Brücke constructed a color ring with six pairs of complementary color hues. He placed seven concentric color hue rings inside the full color hue ring; each ring is one step darker to the center point which is black. Outside the full-color ring, he placed seven concentric color hue rings, each ring one step lighter, the last one white (after 1831, #44 & 46).

Brücke's system is somewhat more logical than Moses Harris's, but it is still not possible here to mix all colors with all grays. It is necessary, in order to achieve this in one color diagram, to add a dimension. First there was the linear presentation with a straight line white-red-black, then a two-dimensional form, a circle (color ring), now the color diagram must become three-dimensional.

13 All Intermediate Colors From Full Color Hue to White, to Black and to all Gray Values

If all color hues, color set at 100, must be mixed with white, with black and with all grays, this can not be accomplished within a flat scheme.

The three-dimensional color diagram

Mutual influences: Mayer 1723-1762, Lambert 1728-1777, Lichtenberg 1742-1799

A solution to this problem is seen for the first time in Lambert's color model, created in 1772. The full colors in this triangular color hue ring, become progressively less colorful towards the middle point of the triangle where color equals 0 at the black point. A perpendicular line is placed in this black point whereupon all the grays are imagined from black, ending in white (#48). Now it is possible to combine all the colors of the triangular ring with the black point. All colors, of this same ring, can also be combined with (color mixtures are shown) all grays from black to white. Thus various pyramidal forms result, with three-, four-, five-, six-, seven-, or other more-cornered or star-shaped surfaces. We can place all these forms into a cone form for easier comparison.

The full color hues are not ordered according to their own intrinsic lightness between black and white

The distance from full color to white and to black are not equal

A cross-section of a pyramid or cone, from the top through the middle to the bottom surface, reveals an unequal distance between colors of the color ring and the white point and the colors of the color ring and the black point. These pyramidal and conical forms are found in the color systems of, among others, Lambert-1772 (#48), von Bezold-1876 (#50), Wundt-1893 (#51). We see this inequality of distances also in the Chevreul's half ball-1839 (#49) and the CIE color system-1931, in the CIE stereo demonstration color model by MacAdam-1935 (#52), Villalobos-1947 (#53), Rösch-1953 (#54), and Richter's DIN 6164 color system-1955 (#55). Some color systems show a curve or deviating angle in the cone base form and/or the cone wall (#49, 53, 54 & 55). The distance from the full color to the white and the distance between full color to black is not equal in the pyramidal and conical forms. A solution to this problem of unequal distances will be shown in the following chapter.

14 Distances From Full Color to White and Black are Equal

Isometric diagrams

A double cone is the result of a search for a spatial model in which the distance between the color ring and the black point is equal to the distance between the color ring and the white point, (#56). There are differences, here also, in the various basic forms such as the triangle, square, multi-angle, star, etc. These forms were used in the color models by: Mayer-1758 (#57), Höfler-1883 (#61), Ebbinghaus-1902 (#62), Ridgeway-1912 (#64), Ostwald-1917 (#65), Boring-1929 (#66), Plochere-1948 (#67)-Hard, NCS-1968 (#68). The equal distance from the full color ring to the white point and to the black point is also realized in the ball form. This form is utilized by Runge-1810 (#58), Wundt-1874 (#59), Munsell-1905 (#63) and Itten-1961, (Runge's color ball) (#58). Finally, the cylinder is used as a basic form by Rood-1879 (#60).

15 Lightest Colors Closer to White— Darkest Colors Closer to Black

Reinstatement of the own intrinsic lightness of full color hues, relative to black and white (Ancient Greeks)

Lightness axis is oblique

Yellow nearer to white Blue-violet nearer to black

If we compare the full colors, color set at 100, with each other, we can determine that in addition to the differences in color hue there are also light/dark differences. Yellow is a pronounced light color and ultra-marine blue, by comparison, a darker color, (Ancient Greeks). Fewer visual color steps, of equal size, are needed from the light color yellow to the white point than to the black point. The reverse is true with the dark ultra-marine blue. Here a greater number of visual color steps, of equal size, are necessary to get to the white point, and fewer to the black point. Titchener places the neutral lightness axis obliquely in order to accommodate these differences in lightness in distance to white and to black, 1887, (#69-70). The white of this neutral axis lies closer to yellow and further away from the darker ultra-marine blue. The black, on the other hand, is closer to the darker ultra-marine blue and further away from the light yellow. The color intervals are of a more equal size.

16 From One Lightest to One Darkest Color

Albert H. Munsell, 1858-1918

The American painter Albert H. Munsell, in 1915, found a logical solution whereby the colors can be ordered according to their own lightness in relation to white and black.

He leaves the neutral axis, with lightness values from white to black, vertical, and places the color ring obliquely (#71). This oblique color ring actually originates through the placement of the full colors at the end of the color arms (#72). These color arms, each of one given lightness, are horizontal; each starts at the neutral axis and ends in a full color hue, i.e. yellow, orange, red, violet, blue and green. Each color arm points to a color hue on the color circle. All colors which lie upon the surface between the lightness axis and the full color point, are of one and the same color hue. The lightness and saturation of each color surface remains a variable color factor. The color arms begin without color at the neutral axis and gradually increase in color to the highest possible color value (intensity) of each given color. The color arms, when they are placed between white and black according to their own intrinsic lightness, will reach their greatest saturation point. If an arm is placed at a higher or lower lightness, less saturations steps of equal size, starting from the neutral axis, will be possible. The light yellow which is near to the white end of the lightness axis, will have a long arm, and on the darker end of the axis, a short arm. The darker color blue-violet has a long arm at the darker end of the neutral axis and a short arm at the light end of the axis (#72 & 123).

Cyan and magenta are missing from the first design (1915)

Munsell (1915) did not include the colors cyan and magenta in the first design of his brilliant color system, There were three main reasons for this:

1. Artists sought more harmony after Newton's irregular color ring. Harmony is increased by creating color sectors of equal size and by a gradual transition between colors next to each other in color hue and lightness. The color cyan had already been replaced (Chapter 10) by a darker blue-green. The color magenta was not present.

2. Maxwell's light-mixing colors, cyan, magenta and yellow, were not regarded as basic colors. Cyan, magenta and yellow lie nearer to the white point, in Maxwell's diagram (#39), than the colors blue, green and red. This would suggest that cyan, magenta and yellow are less saturated than the other colors, which are further away from the white point. Cyan and magenta, originally unknown as paint colors, were regarded respectively as a blue-green adulterated with white and carmine adulterated with white. Yellow was already fully accepted as a saturated basic paint color. Yellow, blue, red and green, as unique colors, are attributed to Leonardo da Vinci (#3). See the chapter 'the unique primeval colors—'.

3. Magenta and cyan were hardly known as painter's colors before 1915, and as printing inks they were not yet geared to each other (together with yellow) as we know them today.

The Munsell system today

The Munsell system is always adapted to new developments

The present day Munsell color system provides space for a greater variety of saturated colors. The system is brought up to date, from time to time, in accordance with new developments. This is possible because of the fundamental design of Albert H. Munsell's color ordering system.

The color ordering (#72), as was originally introduced by Munsell in 1915, from one light color (yellow) to the darker color (violetblue), is also used by Pope-1929 (#73), Johansson 1939 (#74) and Hesselgren-1953 (#75).

17 Three Dark Basic Colors—Three Light Basic Colors

Maxwell's triangular color scheme

Three dark basic colors are diametrically opposite to three light basic colors, in Maxwell's system; he laid the foundation for three-color printing with his color scheme. Transparent colored inks were developed, for three color printing, composed so that each ink color absorbs light mainly in one wave length area (short, middle or long) for which the eye is specially spectrally sensitive.

Color inks with a given absorption area

The color printer Hickethier-1940 (#82), as did Benson-1868 (#80) and Charpentier-1885 (#81), used the cube as the basic form for his spatial color ordering. Charpentier could not yet make use of the attuned printing inks. He is limited in his color choice and therefore mislead as were Itten and Runge. The color pairs which lie opposite to each other in his color diagram are not complementary. His light color orange is closer to the black point and the darker ultra-marine blue is closer to white.

Benson-1868 (additive) and Hickethier-1940 (subtractive) first to show spatial color models with cyan and magenta.

Hickethier could work with the three specially attuned printing inks, developed for three color printing: magenta, yellow and cyan. He designed his thousand-color cube in 1940 (#82). He called the basic colors yellow, magenta and cyan by the names yellow, red and blue, which leads to confusion. These color names are, however, the norm in the printer's world. Hickethier placed the complementary color pairs: white/black, yellow/ultra-marine blue, magenta/green and cyan blue/red, at diametrically opposing corner points in his color cube.

Confusing color names.

He gave the colors ultra-marine blue, green and red the names violet, green and orange. He placed the three basic printing ink colors yellow, magenta and cyan on the three corner points around the white point. The darker colors, red, green and ultra-marine blue, each the result of printing two lighter basic printing colors one over the other, are placed on the three corner points around the black point. Three lighter colors are near to white, and three darker colors near to black in this diagram (#77-79 & 82).

If we make a cross-cut through Hickethier's cube over the three basic mixing colors of three-color printing, yellow, magenta and cyan, the result is an equilateral triangle. An equilateral triangle is also visible if we make a cross-cut over the colors ultra-marine blue, green and red. Here all six main colors lie equidistant from the neutral point, which coincides with the lightness axis, unlike Maxwell's triangle.

The six basic colors equidistant from the neutral axis

The lines interconnecting the basic colors of the top triangle are straight, as well as those between the basic colors in the bottom triangle. The mixed colors on one side of the triangle, though, are much closer to the neutral of the lightness axis. None of the mixed colors, however, is neutralized in the

All mixed colors (color = 100%) nevertheless, lie closer to the neutral axis.

least by a complementary color which lies opposite to it. All colors, color set at 100, are placed on the cylinder wall surrounding the central neutral axis, each full color is equidistant from this axis. All colors therefore, color set at 100, do not lie on a straight line between two basic colors, but on a cylinder wall surrounding the central neutral axis. There is no place for 100% saturated mixed colors in the cube or rhombohedron (rhombohedron = space enclosed by six diamond-shaped surfaces, rather than a space enclosed by six square surfaces = a cube). A cube or rhombohedron is a useful model for placement of a scheme for mixing properties of printing ink; ink mixing colors can never be 100% saturated. The cube and the rhombohedron are not useful as a color space because they can not enclose all saturated mixed perception colors.

No place for 100% saturated mixed colors

Three different lightnesses on one and the same lightness level

There is another fundamental objection to the use of a cube or rhombohedron as a color space: the basic colors can not be placed according to their own intrinsic lightness in relation to the lightness axis. Yellow, magenta and cyan, each with their own different lightness, are placed on one and the same level of lightness. Ultra-marine blue, green and red, also each with its own different lightness, are placed on one and the same lightness level. None of the basic colors and/or mixed colors in the cube or rhombohedron, can be located logically as to its own inherent lightness. Küppers, 1972, made the same color ordering as Hickethier's. He used an elongated cube, a rhombohedron, as his basic form (#83).

No color is logically located according to its own inherent lightness.

Rhombohedron

Lightness relationships are disoriented

The rhombohedron has the same shortcomings as the cube when used as a basis for a color model. The lightness relationships are distorted even further by the elongation of Hickethier's cube. The color lightness relationship between the colors yellow and magenta, as compared to that of magenta and red, are even more disoriented than in Hickethier's cube.

The color-perception-space Gerritsen, 1975 (#84-88 & 124) will demonstrate to us how all perceptible colors can be placed. All colors, in this color-perception-space, are ordered according to the laws of color perception as to their own lightness, color hue and saturation.

18 The Color-Perception-Space, Gerritsen — 1975

The construction of the color-perception-space

First it is necessary to consider a schematic summery of three factors in the process of vision: light, object and eye; this in order to better understand the construction of the schematic color-perception-space.

Light—object—eye

Light (energy)

Light (actually light energy): is the only energy source to give us information from a distance for viewing our perceptible environment.

Object (absorption/reflection)

Object: is the perceived matter which, selectively, absorbs part of the light energy received; the rest of the light energy, received in the direction

Eye (perception)

of the eye, gives us information about the object. This remaining light energy is called the reflection. Light which passes through transparent materials is called transmission.

The part of a remission or reflection that reaches the eye gives us visual information about the object.

Eye: color and lightness perception possibilities of our visual sense (eye, brain, consciousness) in relation to the received light energy.

The connection between the three factors of the sight process: light, object, eye, is shown schematically in the color perception schemes we have developed.

The color perception scheme

Part of light—object—eye can be shown

The information regarding these three factors will be presented schematically. The schemes representing this information have been attuned to each other; they show which part light(energy), object and eye each have in the perception of color (and lightness). We call these schemes color perception schemes. We will use these color perception schemes to better understand the construction of the color-perception-space. We will place one color perception scheme per color in each phase of construction.

A simplified scheme, as part of the color perception schemes, will be presented at the end of each of the following sections: light, object and eye.

Light

Light energy gives the eye information

Light is a visual sensation, caused by the light energy. It is an electromagnetic radiation in a wavelength area within which our eye is capable of receiving energy and 'translating' it into visual information. The wavelength area lies between the wavelengths of 380 and 760 nm = nanometers = millionths of a millimeter. Energy of somewhat longer wavelengths gives the eye no visual information, but rather heat information. Somewhat shorter wavelengths could injure the eye.

Measuring the daylight spectrum

Measuring light energy

We can measure the amount of energy per wavelength area with a spectro-radiometer. A schematic presentation of the measurement of daylight under an overcast sky in small wavelength areas is found in #90.

Standardized light sources

Different kinds of light

There are four internationally accepted standards of light: A, B, C and D, each with an established energy curve (#91). We draw a horizontal line through the intersection point of the four curves; the horizontal line is dotted in the illustration. This dotted line represents the upper limit of our simplified scheme for light: light energy 1, in all wavelength areas = 'white'; the lower edge of our scheme represents light energy 0, in all wavelength areas = 'black' (#91 below & 95).

Scheme for light

Scheme for light energy

Our scheme for light is a rectangle, left represents the short wave area and right the long wave area of the daylight spectrum; the top edge is white in all wave length areas and the bottom is black in all wavelength areas (#95).

Object
Asborption by the object

The curve will look approximately as shown in #103 A, if we hold a piece of transparent red glass in front of the small light slit while measuring the daylight spectrum (#90). Light (energy) is let through only in the long wave area to the right; The rest is absorbed by the red glass and transformed into heat. The curve of this red light can come from a red light source as described above, or it can also be the remaining light that is reflected by a red surface (#92).

The reflection curves of the six basic colors are shown in #99-A through 104-A. We have simplified these curves, because they are difficult to read, and incorporated them into the light scheme (#99-B to 104-B).

Scheme for object

We can now see clearly in which wavelength area light energy of the light scheme is mainly let through, 'reflected', or held back, 'absorbed', in the short-, middle- or long wave areas of the light spectrum (#96). The importance of the light scheme becomes clear. If the light source is not standard light, but a sodium lamp which contains no short wave light, a blue glass, which lets short wave light through, will receive no light from the short wave area which it can let through to the eye. The blue will look black or colorless. We will use our completely stylized standard light scheme in the construction of the color-perception-space (#95).

The eye

The eye and color perception possibilities

If the eye receives mainly short wave light (#104-A to E), an ultra-marine blue color sensation results. Light mainly from the middle wave area of the daylight spectrum results in emerald green (#102-A to E). Light from the long wave area is experienced as red (#103-A to E).

International standard CIE (Commission International de l'Eclairage)

The spectral distributions CIE 1964 (spectral tristimulus values) are shown in #93. They have been accepted internationally as the standard for color measurement with a 10 (ten degree) field. One must imagine three light energy receptors (photo cells) in which the relative spectral sensitivities correspond to the CIE 1964 spectral distributions, instead of the color sensitivity of the eye.

Scheme for the eye

We have schematized this information for our color perception schemes. Spots in colors which correspond to the color sensation which is summoned by light energy in a given wavelength area: short-blue, middle-green, long-red.

Our scheme for 'eye' consists of the blue, green and red dots (94 & 97).

Composition of the color perception schemes

The schemes for the type of light, absorption/reflection by the object and the visual color response by the eye, are combined in one color perception scheme.

The object absorption-(reflection-/transmission-)scheme (#95 & 96) is of the same size as the rectangular light scheme under which it is placed (#95). The scheme for color response by the eye (registered spectral range, #96 & 97, together form #98) is placed under the other two schemes. We can come to understand different color mixing properties with these color perception schemes. This applies to additive, subtractive and partitive color mixing or formation. See also chapter 3: 'The visual color image'.

Measurement of light energy minus absorption is reflection

Absorption/reflection curve is stylized—significance of the light scheme

Relation between light wavelength and color response

CIE standard observer

The scheme for color perception

Light Absorption/reflection Perception

The main divisions of the color perception space will be explained as to color hue, lightness and saturation, in the following chapters, 19, 20 and 21.

19 Color Perception and Color Hue

Additive

**Construction of the color circle, additive
(with color perception schemes for each sector, p. 68)**

We start with a white projection screen (a screen which reflects all light wavelengths) in a dark room. The white projection screen looks black (#105). We will project a color circle with colored light, corresponding to the CIE standard wavelength areas, onto this screen. The circle is divided into six sectors. Outside the circle, the color perception scheme for each sector is shown; these schemes indicate in which wavelength area light is mainly reflected and what the color response will be.

We project light energy in the short wave area of the light spectrum onto three adjoining sectors (half a circle) (#106). Perception for short wave light is activated; the color sensation is ultra-marine blue.

Color perception, short wave area: ultra-marine blue

We can 'read' the (three) color perception schemes by each of the three blue sectors; the receptors for short wave light are activated; the color sensation is ultra-marine blue.

Color perception, middle wave area: emerald green

We now project light energy in the middle wave area of the light spectrum onto three adjoining sectors (#107). Perception for middle wave light is activated; the three color perception schemes show us the color sensation is green. Where the projections of short and middle wave light overlap in one sector (#108), the color cyan appears. This color always appears when light in both the short and middle wave areas is equally and simultaneously activated. Since two types of receptors are simultaneously activated in one color sector, the resulting mixed color is lighter than either of the two original (receptor) colors blue or green. The color perception schemes show clearly in which spectral areas the color receptors are activated.

The additive mixed color of blue and green is cyan

We now project light energy in the long wave area of the light spectrum onto three adjoining sectors (#109). Perception for long wave light is activated; the color perception scheme shows us the color sensation red. Where the projections of middle and long wave light overlap in one sector, the color yellow appears (#110); complementary to ultra-marine blue. This we can also 'read' on the color perception schemes. When the receptors sensitive for middle wave light (green) and for long wave light (red), are activated equally and simultaneously, the color sensation yellow appears. The mixed color yellow is also significantly lighter than either of the mixing colors, red or green.

Color perception, long wave area: red

The additive mixed color of blue and red is magenta

Where long wave light (red) and short wave light (blue) overlap, the color magenta appears (#110). The mixed color is lighter than either of the mixing colors, ultra-marine blue or red. The color magenta is not present in the fanned-out daylight spectrum, the blue and red, which together make magenta, are at the two opposite ends of the spectrum. The spectral sensitivities of the eye do not overlap each other enough. The overlapping of spectral sensitivities is clearly seen in the spectral colors yellow (middle + long) and cyan (short + middle). If we look at the color perception scheme for the magenta sector, it becomes clear that the color perceptions for both

The additive mixed color of green and red is yellow

58

The additive mixed color is lighter than either one of the mixing colors

non-overlapping spectral color sensitivities can be activated equally and simultaneously in the eye.

Complementary color pairs opposite each other

If we compare the color perception schemes of the colors lying opposite to each other, we can see that these color pairs are complementary. Where these colors overlap each other in the center of the circle, a white triangle appears. The color perception schemes of two color sectors opposite each other show that the sum of the activated color receptors is always three.

The additive mixed color of a complementary color pair is white

The color receptors sensitive to short-, middle- and long wave light are activated equally and simultaneously when the colors of a complementary color pair overlap; the color sensation white appears.

Light energy in mainly only one wavelength area, short-, middle- or long-, is projected onto three circle sectors at a time in the additive color circle construction. We will start again with the same circle, which is white, for the subtractive construction of the color circle; we will now subtract light energy in the direction of the eye in only one of the wavelength areas at a time, short, middle or long, from three circle sectors.

Construction of the color hue circle, subtractive (With a color perception scheme for each sector, p. 69)

Subtractive

We start with a white surface (#111), big enough for the color circle and the six color perception schemes. The three color perception symbols for ultra-marine blue, emerald green and red, are placed beside each circle sector as a color perception scheme for white.

Light energy mainly in one wavelength area is absorbed, or transformed into heat, by the colored transparent ink or paint on half of the circle (three adjacent sectors). The color perception schemes, which correspond to each circle sector, show in what wavelength area light energy is absorbed and in what wavelength area light energy is let through in the direction of the eye. The yellow ink absorbs mainly short wave light.

Color perception minus short wave area: yellow

We can see, on the color perception schemes, that the receptors sensitive for short wave light can no longer be activated (#112). The color perception sensitive for + middle and + long wave light are simultaneously activated, or the 'minus short' color perception, the color sensation yellow appears. No light of any wavelength has as yet been absorbed in the three other color sectors. Here the color receptors sensitive to short-, middle- and long wave light remain activated simultaneously; the color sensation is white.

We now apply magenta ink to three sectors (#113). This ink mainly absorbs light energy from the middle wave area of the daylight spectrum. The color perception schemes show us that only the receptors, sensitive for short- and long wave light simultaneously, are activated; the rest of the light is absorbed by the ink. Light is subtracted twice by absorption in the overlapping sector, by two layers of ink each in a different wave length area.

Color perception minus middle wave area: magenta

Only in the long wave area is light still let through in the direction of the eye. The color receptors sensitive to long wave light alone are activated. The color sensation is red (#114).

The subtractive mixed color of yellow and magenta is red

The ink for the three remaining sectors is cyan colored (#115). It absorbs the light in the long wave area of the spectrum. Where the cyan colored ink is printed over a yellow colored ink, the color green appears (#116). The yellow ink absorbs in the short wave area, the cyan colored ink removes light in the long wave area; light is still reflected in the direction of the eye

Color perception minus long wave area: cyan

The subtractive mixed color of yellow and cyan is green.

The subtractive mixed color of yellow and cyan is green. The subtractive mixed color of magenta and cyan is ultra-marine blue.

The basic color circles additive and subtractive are the same

The subtractive mixed color is always darker than either of the two component colors.

The mixed color of inks has the average lightness of the two components when two colors are mixed.

Color circle with 12 color hues

Color circle with 18 color hues

The infinitely finely divided color hue ring

only in the middle wave area. Only the receptors that are sensitive to middle wave light are activated. The color sensation is green.

Where cyan colored ink lies over the magenta, the mixed color ultra-marine blue appears (#116). Magenta ink absorbs light in the middle wave area; cyan ink absorbs light in the long wave area. Only light in the short wave area is not absorbed and is reflected in the direction of the eye. The receptors sensitive for short wave light are activated; the color sensation is ultra-marine blue.

Here we have a complete subtractivly constructed color circle with a color perception scheme per sector. If we compare the subtractively constructed color circle with the additive one, we can determine that the color sectors as well as the color perception schemes of both circles are the same. The only difference is that where the colors overlap in the center of the subtractive circle, a black triangle appears, instead of the white triangle where the additive colors overlap.

The basic colors are darker then the component colors, from which they are mixed when two transparent colors overlap in one sector.

Ink mixing

The mixed color of inks has the average lightness of the two components when two colors are mixed. The darker color appears to be diluted by the lighter color.

The subdivision of the six basic color hues

A visual mixed color hue is placed between each of the six basic color hues, a color ring of twelve color hues is formed. Orange lies between yellow and red, yellow-green between yellow and green, blue-green between cyan and green, etc. (#43).

If, instead of one intermediate color hue, we place two equal color hue steps between each of the six basic colors, an eighteen-part color circle results (#117). The color perception schemes, placed beside each sector, show how much and to what extent, and in mainly which wavelength areas color perception is activated. Light in mainly one of the three wave length areas, short, middle or long wave, activates the sensitive color receptors with three basic color hues. These are the basic color hues: ultra-marine blue, green and red. We could name these the three eye primaries. The other three basic color hues: yellow, cyan and magenta, mainly activate the sensitive color perception for two of the three wavelength areas, short, middle or long; for: yellow = middle + long, cyan = short + middle, magenta = short + long. We could name these three colors the special eye secondary colors. An infinite number of eye secondary colors are possible between the eye primaries and the special eye secondaries. Together they form a finely divided color hue ring (#117). All color hues here have a color value of color = 100, neutral = 0, because none of the color hues is at all neutralized by a complementary color lying opposite to it.

20 Color Perception and Lightness

The neutral lightness axis
(with color perception schemes, #118)

Color perception neutral as black

The color perception schemes at the start of the additive or subtractive construction of the color circle with the six basic color hues, show that there is a neutral balance between the mutual color perceptions for short-, middle- and long wave areas of the spectrum. We see that none of the three wavelength areas receive light energy at the start of additive construction #105. None of the three color receptor groups can be activated with a color sensation, only a neutral black. All three wave length areas receive light energy at the start of the construction of the subtractive color circle #111. All three types of color receptors are equally and optimally activated. The color sensation which appears is white, a neutral, only lightness, as in the description of a color without color hue.

Color perception neutral as white

There is no color hue because all three eye primaries are equally activated so that none of the three color sensations blue, green or red can dominate. White has the greatest lightness, as a color surface with the highest percentage of reflection, and also the greatest lightness, as perception as compared to other lightnesses. If we activate the receptors equally but not optimally, e.g. to only half of our lightness adaptation potential, we will see gray instead of white. Gray is also neutral, a color without a color hue. A schematic lightness axis, with color perception schemes is presented in #118. All these colors without a color hue, from black to white, are called the special tertiaries.

Color perception neutral as the grays

A practical example of special tertiary colors is a black and white program received on a color television set; the entire black and white picture is made up of mutually equally activated phosphor spots in ultra-marine blue, green and red. The general impression is, however, of a black and white image with many gradations of gray. Such a series of lightnesses without color, from white via many grays to black, is usually called a lightness axis or neutral axis, as part of a color diagram.

Lightness axis

The inherent lightness of the basic color perception hues

#117 illustrates an infinitely finely divided color hue ring. The hypothetical primaries ultra-marine blue, green and red occupy three points on this color hue ring. Ultra-marine blue has the lowest inherent lightness, then comes red with a somewhat higher inherent lightness and then green with the highest inherent lightness of the hypothetical eye primaries. The eye secondaries yellow, magenta and cyan lie in between the hypothetical eye primaries. The inherent lightness of magenta is lighter than the lightest eye primary color, green; cyan has a higher inherent lightness than magenta and yellow and the highest inherent lightness of the six basic colors.

The zigzag line of eye primaries and special eye secondaries

We see that the hypothetical eye primaries and the special eye secondaries alternate in the basic color hue circle. The color perception schemes also make this clear. We imagine that the lightness axis is placed vertically through the middle point of the basic color hue circle; the white point is at the top and the black point at the bottom of this axis. The lightnesses are coded from white to black with the letters A through J. We can place each basic color hue according to the same lightness as is found on the lightness axis. The color hues now no longer lie on one level of the color circle, but on the outside of a cylinder wall with the diameter of the color circle.

Inherent lightness of all eye primaries and secondaries

The own intrinsic lightness of the basic color perception hues presented schematically: yellow -C, red -G, magenta -E, ultra-marine blue -H, cyan -D, green -F, and then again yellow -C (#124).

If we link all these color points with a line on the cylinder wall, an irregular zigzag line results (#124). Color = 100 (the infinitely finely divided color hue circle) of all colors can be found on this irregular zigzag line. These full colors are now all ordered as to color hue and their own inherent lightness according to the laws of our lightness and color perception.

21 Color Perception and Saturation (Colorfulness)

The saturation of the colors (with color perception schemes—#123)

All imaginable eye primaries and eye secondaries, the saturated colors, color to be set at 100, now have their places on the cylinder wall in the irregular zigzag line, -E -H -D -F -(C).

The special eye tertiaries, white via all gray gradations to black, color to be set at 0, together make up the lightness axis of the cylinder with the schematic lightness divisions from A to J. A vertical cylinder cross-section is made in order to show the schematic progression from a color, color is set at 100, on the cylinder wall to a neutral, color is set at 0 on the neutral lightness axis (#123). This cross-cut through the complementary color pair ultra-marine blue and yellow shows us the following: all the blues here have one color hue, blue; all the yellows have one color hue, yellow. All the saturation gradations of a color have the same lightness on each horizontal line. The color perception schemes on the cross-cut illustration schematically show us the differences in saturation levels.

Saturation gradations ultra-marine blue/yellow

If we trace the saturation gradations from full color hue yellow to a neutral of the same lightness, we see that in the first color perception scheme, mainly the color perception sensitive for energy in the middle- and long wave area is activated.

The yellow perception is neutralized by the complementary blue perception in the neutral axis. It is not only the influence of the perception sensitive to energy in the short wave area of the spectrum which has increased. As the short wave area's influence becomes greater, the middle and long wave area's influence becomes proportionally less great. If this was not true, then the total amount of energy would have increased. The result would be that the lightness degree would not have remained the same and the yellow would look lighter. It would have changed from yellow via cream towards the white point.

We see, in the color perception schemes, that as the blue perception increases, on the top side, green + red perception = yellow perception is reduced proportionately. We can clearly see the concept of the inherent lightness of a color hue, color is set at 100, demonstrated in the schematically presented yellows. If the inherent lightness of the yellow is increased, the yellow becomes less yellow, lighter; the saturation decreases. If the lightness decreases, the yellow changes towards dusty ocher via green brown to dark bronze.

Color hue and highest saturation degree determine the inherent lightness.

Yellow only on one lightness level

Blue on several lightness levels

Ultra-marine blue, which has much less inherent lightness than yellow, is governed by the same principles. As lightness increases, saturation decreases. The blue then changes from deep blue to somewhat lighter blue, grayer blue and via pastel blue toward the white point. If the lightness is lower than the inherent lightness of the blue, the saturation also decreases. The deep blue then changes via midnight blue and blue-black to the black point, where the saturation is set at 0.

A typical difference between yellows and blues is that yellows are actually yellow only at one point; the blues from light to dark blue, on the other hand, still belong to and are recognized as blues. This characteristic is illustrated in Johansson's (1939; #74) color diagram, where the yellow (gul) lies only on one lightness level, while blue (blå-blå) appears on different lightness levels.

One can also keep the color hue and lightness of blue constant and decrease the saturation only. Here, too an increased influence of the perception in the middle and long wave area (complementary color pairs), and a proportional decrease of influence from the short wave area follows if lightness is kept equal. Naturally yellow and blue can become less saturated according to all the possible connecting lines with any given point on the lightness axis, from white to black. It is also possible to place a line straight through the color perception space at any given angle so that color hue, lightness and saturation change.

The same principles apply to all color hues, color set at 100, of the infinitely divided hypothetical outer color hue ring (#117).

22 The Three-Dimensional Schematic Color Perception Diagram

All the points of this three-dimensional color hue ring are linked above with the white point and below with the black point. These lines form the outer limits of the color perception space.

Optimal color hue to white— optimal color hue to black

Here the color hues succeed each other from full color, color set at 100, to the white point, color set at 0. (optimal colors of increasing lightness). All colors also progress to the black point, from color set at l00 to color set at 0 in the black point, (optimal colors of decreasing lightness) (#124).

The color hues inside the color perception space form a fan of color hue surfaces around the neutral axis, and end on the outer contours of the perception space. The lightness levels are projected horizontally at every level between white and black. These horizontal surfaces are bound by the lines of the color hue ring on the cylinder wall to the white and black points (#124). The saturation gradations are projected as an infinite number of concentric cylinder walls surrounding the neutral lightness axis. Saturation equals zero at the neutral axis. The further a color is from the axis, the greater is its saturation within the color perception space (#123).

23 Color Cross-Sections of the Color Perception Space

Cross-sections of the color perception diagram, schematic in color

Six vertical and six horizontal schematic color cross-sections of the color perception space are shown in #125-136.

Vertical cross-sections

The first third and fifth vertical cross-sections illustrate the position of the six basic colors, respectively the complementary color pairs: ultra-marine blue/yellow, cyan/red, and green/magenta. The second, fourth and sixth vertical cross-sections illustrate the positions of the colors lying between the six basic colors

Turned 30

The color space is turned 30 on its axis for each of these vertical cross-sections in order to show the various color hue surfaces.

Horizontal cross-sections; positions predetermined

The color hue fan remains fixed in one position for the six horizontal cross-sections. The position of ultra-marine blue/yellow remains constant on all horizontal cross-sections from C to H. The lightness only changes, as a factor of perception, from one cross-section to another. We have printed the names of the six **BASIC** colors in **BOLDFACE** in the text below, in order to make them easier to recognize. **YELLOW** alone appears in lightness level C as a saturated color. It lies furthest away from the neutral axis. We find the saturated color hues **CYAN** and yellow-green on the horizontal cross-section D (#128). The saturated color hues orange, blue-green and **MAGENTA** at lightness level E. We find light green-blue, carmine red and **GREEN** on cross-section F. The color hues **RED** and violet have their places on cross-section G. Finally, on cross-section H there is only the saturated color hue **ULTRA-MARINE BLUE**. If the color hue ring were divided much more finely, an infinite number of surfaces of the color hue fan would result.

Infinite fan of color hue surfaces

If the visual difference in saturation gradations is lessened, the number of saturation gradations will increase, from the neutral axis to the most saturated color. Then the color hue arms could also become proportionally longer.

If the relative visible lightness level differences are made smaller, (more lightness gradations between black and white) the lightness axis will become proportionately longer. The colors are indicated only schematically, as is stated in the title above.

It would be very difficult to reproduce the colors accurately, with narrow color tolerances, in color printing. We are concerned here only with the schematic presentation of color perception possibilities. Much higher and lower lightness values would be necessary than would ever be possible on paper and with one light source. The saturated color hues would include vividly colored signal lights, the color perception of neon advertisements against the midnight blue sky, the phosphor spots of color television, the deep colors of velvet, the colors of colored glass in all thicknesses with light falling onto and through it....

Schematic presentation of all colors

The schematic colored cross-sections give us an insight as to the place of the various perception colors in relationship to each other. One sees which colors belong together and which are complementary or neutralizing to each other (with simultaneous contrast of larger color surfaces, complementary colors will enhance or strengthen contrast). If we want to present the color perception space more clearly, a construction of the color perception space can be most helpful. The instructions for building one's own color perception model are given in Chapter 25.

(Continued on page 81)

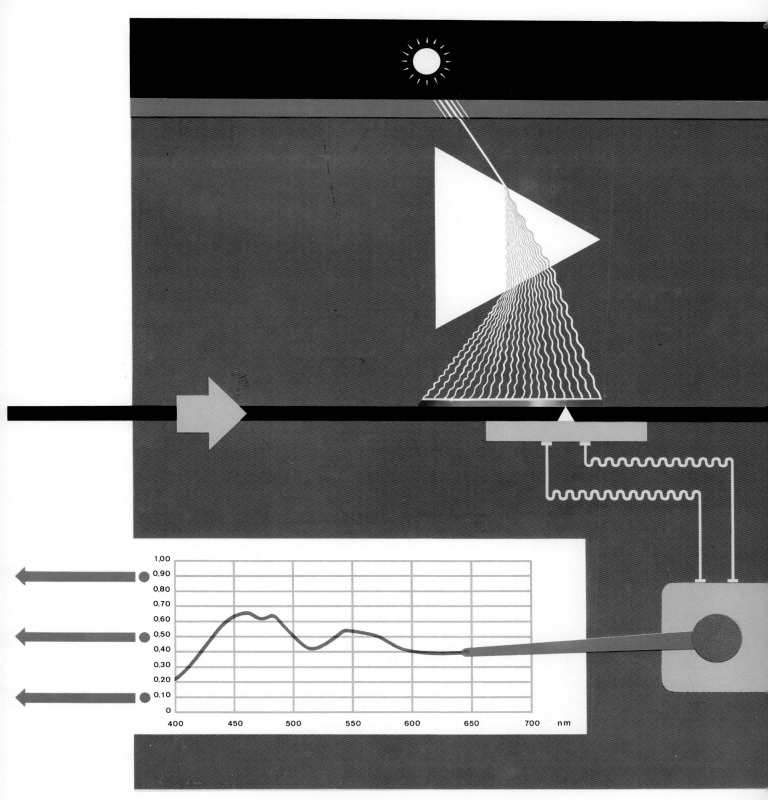

#90 Schematic presentation of measuring the daylight spectrum

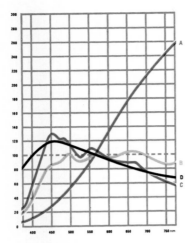

#91 Curves of standard light sources: A = incandescent lamp light, B = sunlight, C = daylight, D = corrected daylight

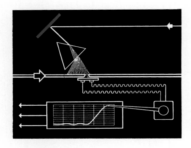

#92 Schematic measurement of the reflection of a red color surface

#93 The spectral distributions CIE 1964 are internationally accepted as the standard for color measurement of a 10 (ten degree) field (the CIE 1964 standard colorimetric observer)

#94 Our scheme for the eye consists of the blue, green and red spots which symbolize the colors which appear, respectively, when light energy from predominantly short, middle and long wave length areas of the daylight spectrum is presented

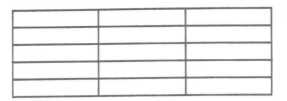

#95 The scheme for light, upper edge light energy: 1 = white, lower edge light energy: 0 = black

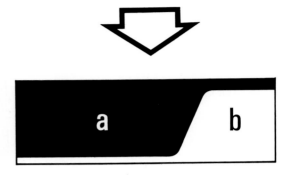

#96 The scheme for absorption/reflection by object: light #95, minus absorption 'a' by the object = reflection 'b'

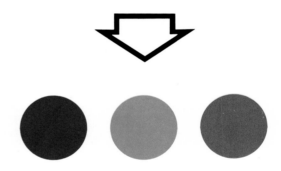

#97 The scheme for the spectral sensitivity of the eye to the same scale as the scheme for reflection 'b'

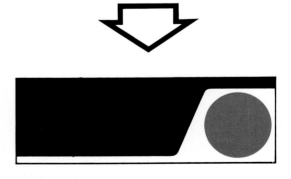

#98 The eye spectral sensitivity sensitive to light in the corresponding wave length area of the reflection of 'b' #96. The activated color sensation is red.

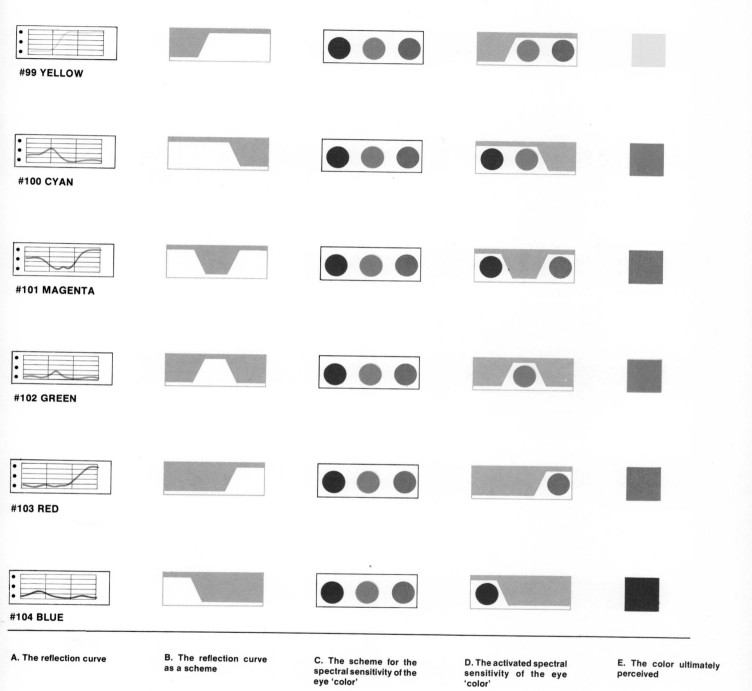

The ADDITIVE construction of the color circle through the projection of respectively short, middle or long wave light energy added in the direction of the eye by these projections which overlap each other. The mixed color is always lighter than either of the colors used to mix it.

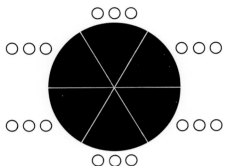

#105 A 'white' projection surface which receives no light. No color perception receptors are activated.

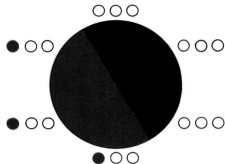

#106 Light from the short wavelength area is projected onto three sectors. The spectral sensitivity for short wave light is activated. The perceived color is ultra-marine blue.

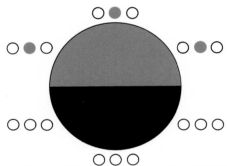

#107 Light from middle wavelength area is projected onto three sectors. The color spectral sensitivity for middle wave light is activated. The perceived color is green.

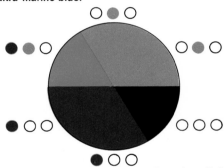

#108 Projections #106 and #107, where they overlap, show light of short and middle wavelengths together. Two kinds of color receptors are activated simultaneously. The perceived color is cyan.

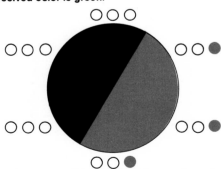

#109 Light from the long wavelength area is projected onto three sectors. The spectral sensitivity for long wave light is activated. The perceived color is red.

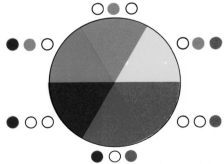

#110 The projections #108 and #109 are combined, the complete color circle is formed. Yellow appears where red and green perception overlap, magenta from a blue and red perception.

The SUBTRACTIVE construction of the color circle, in which light energy is subtracted in the direction of the eye in, respectively, short, middle and long wavelength areas, through placement of transparent ink layers, over each other. The resulting mixed color of two transparent layers of colors is always darker than either of the two mixing colors.

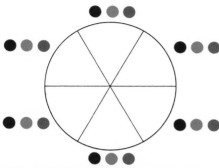

#111 A white projection surface where light is received. All receptors sensitive to three spectral areas are activated. The color perceived is white.

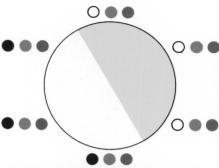

#112 Light energy from the short wave area is absorbed in three sectors. Only green and red sensitivities are still activated as we can see on the color perception schemes. The color perceived is yellow.

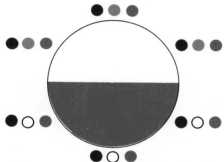

#113 Light energy from the middle wave area is absorbed in three sectors. Only blue and red sensitivities are still activated. The color perceived is magenta.

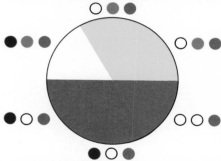

#114 Where yellow and magenta overlap (#112 & #113), only light from the long wavelength area is let through. The color perceived is red.

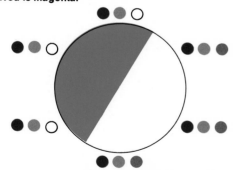

#115 Light energy from the long wave area is absorbed in three sectors. Only green and blue sensitivities are still activated. The color perceived is cyan.

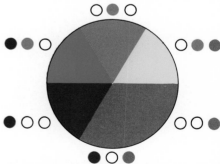

#116 The absorptions of #114 and #115 combined and overlapping, now give us the complete color circle. Cyan over yellow lets only middle wave light through in the direction of the eye, we see: green. Cyan over magenta lets only short wave area light through, we see: ultra-marine blue.

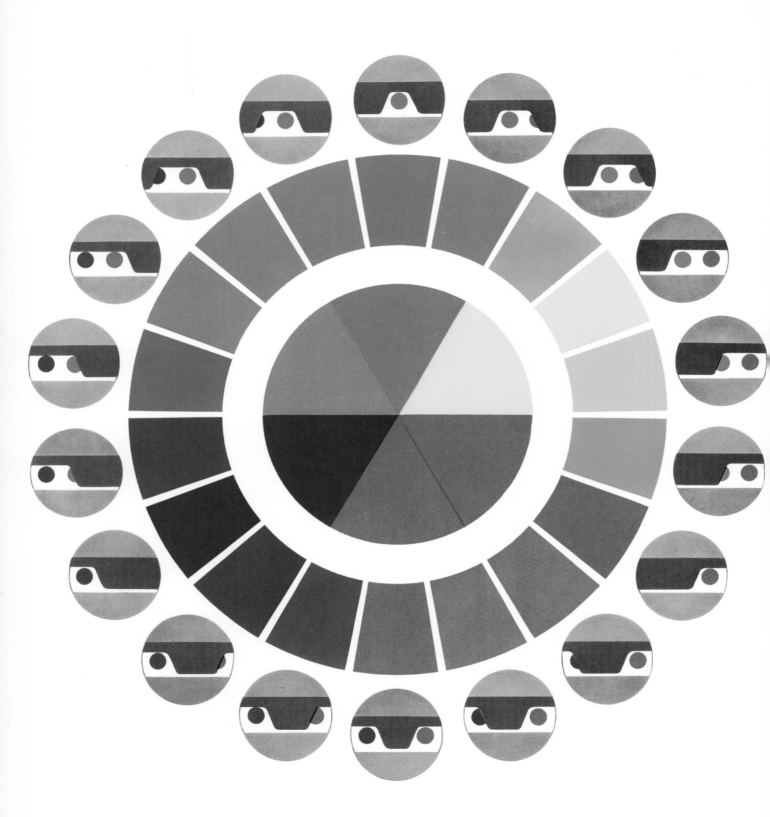

#117 The color circle with further divisions of the six basic colors, with color perception schemes per color sector. One can imagine the infinitely divided color hue ring, of successive colors which flow into each other without visible borders, surrounding this color circle.

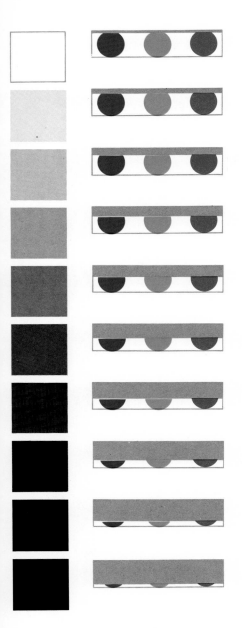

#118 The neutral lightness axis with the color perception schemes. None of the activated spectral sensitivities dominate in any wavelength areas.

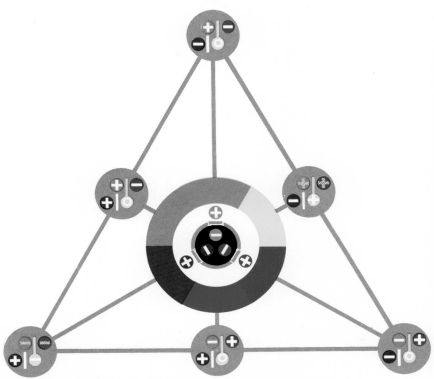

#119 The opponent values as signal transfer in (color) vision, schematic. The blue-yellow and the green-red opponent values are presented schematically on the contour of the triangle. The lines which connect with the center indicate the relationship between signal transfer and perceived color. The blue-green-red circuit is short-circuited by the neutral contrast signal between the white ring and the black center.

#120 First development phase of vision, only dark/light information. The opponent signal, black-white.

#121 Second development phase, from the contrast formation black-white, the spectral perception to the opponent signal blue-yellow develops, with the neutral black-white contrast signal as a reference point.

#122 Third development phase, the process repeats itself: the spectral sensitivity for the opponent signal green-red develops from the yellow information with yellow as the reference point, as the opponent of blue; with the neutral as a reference point of the black-white opponent signal without color hue.

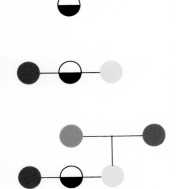

#123 Schematic cross-section of the color perception space through the complementary color pair ultra-marine blue/yellow, with color perception schemes for each color surface.

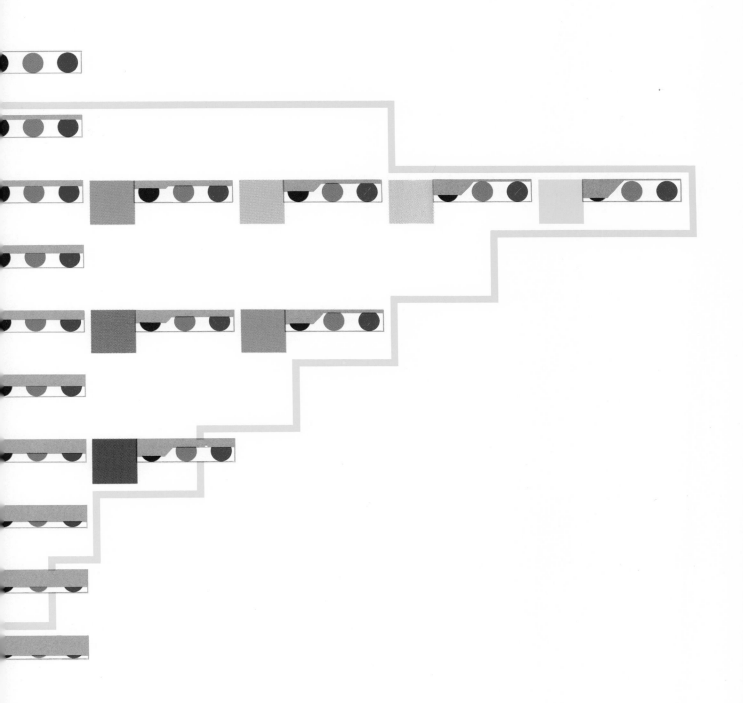

Comparison with Johansson's-1939 color diagram, #74. Yellow can only be called real yellow on one light/dark value; blue, from light to dark, remains blue. We see only one point for yellow (gul), but a much larger area for the blue (blå-blå), in Johansson's color diagram.

#124 The three-dimensional color perception diagram: color hue-lightness-and saturation-perception with eight color perception schemes, Gerritsen, 1975

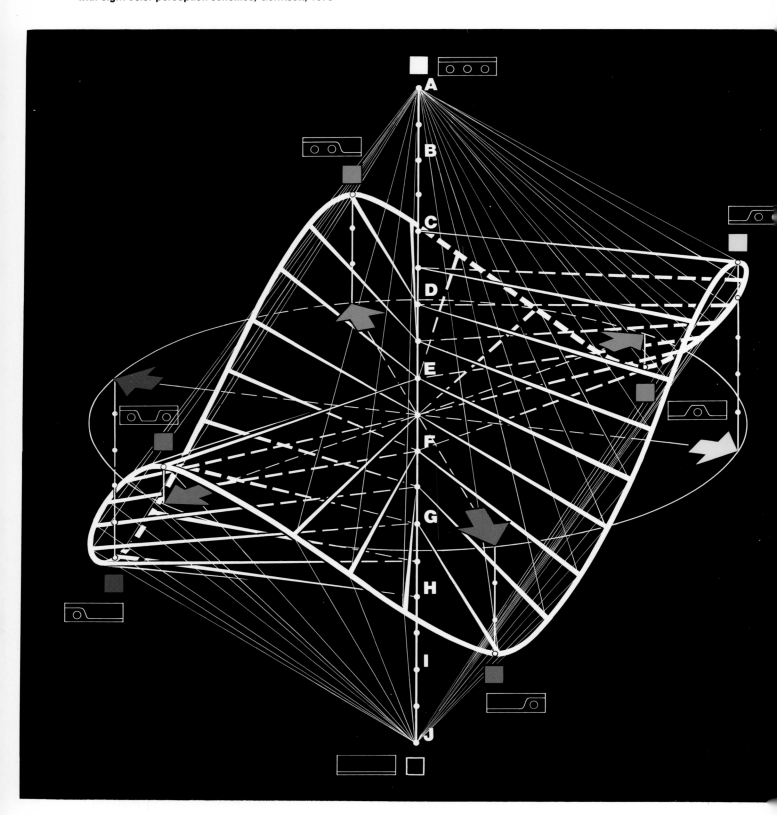

#125

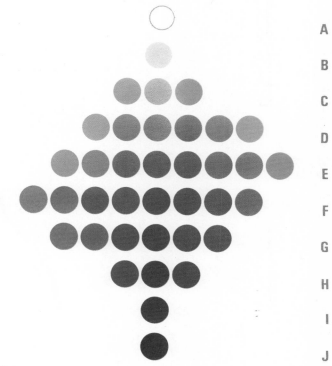

#126

A
B
C
D
E
F
G
H
I
J

Schematic in color: six vertical and six horizontal cross-sections through the three-dimensional color perception diagram on page 74 (#125 through #136).

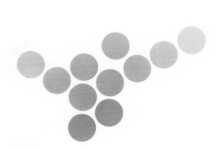

#127

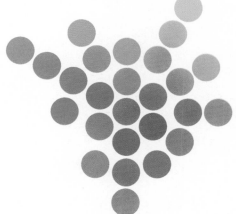

#128

C

D

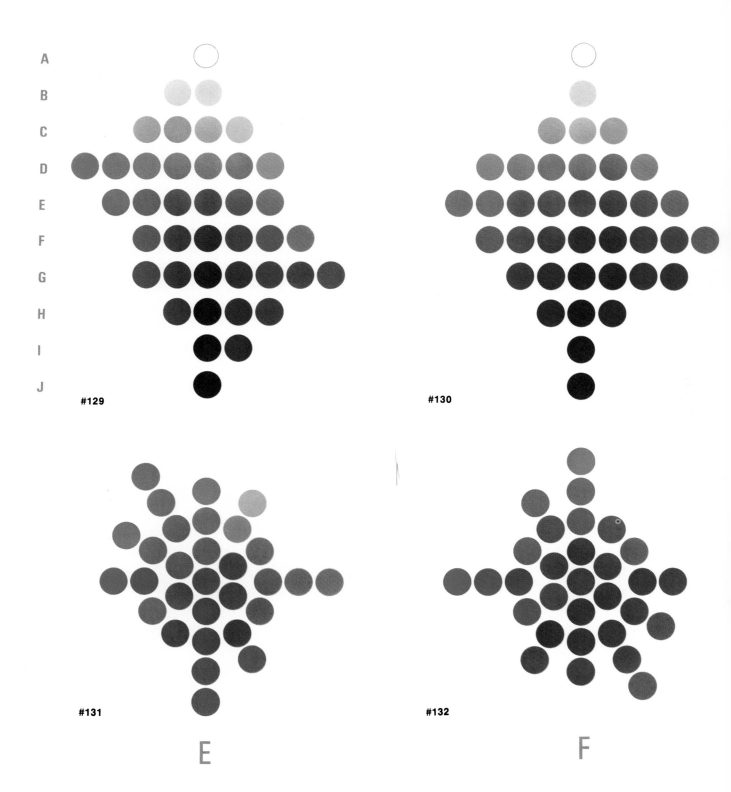

#133

#134

#135

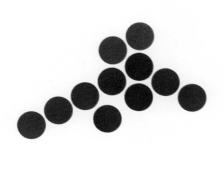

#136

G H

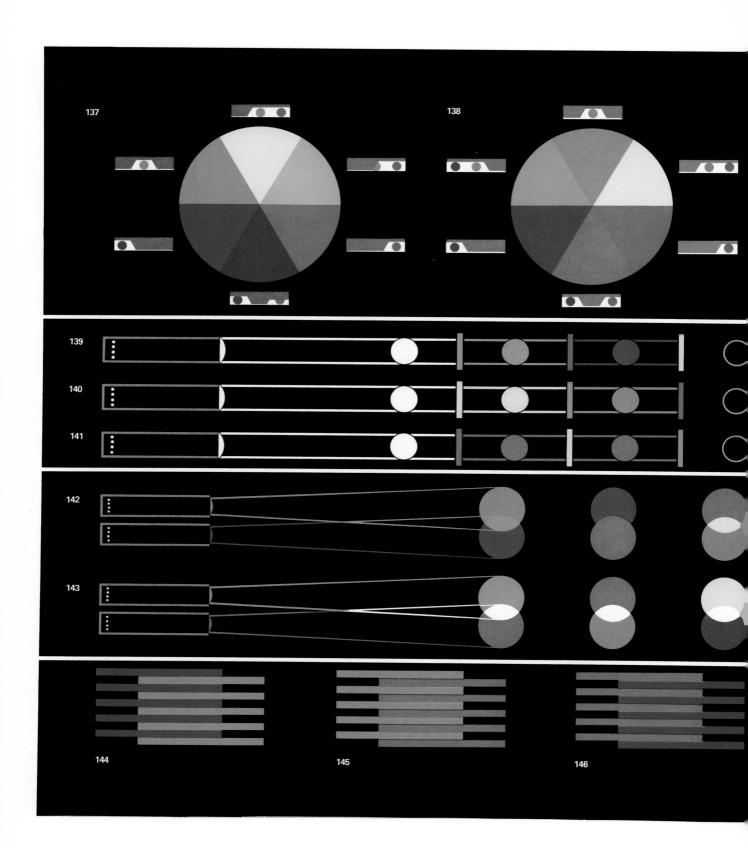

UNDER THE LINE HERE AND CUT WITH A SHARP KNIFE ALONG A METAL RULER

14	13	12	11	10	9	8	7	6	5	4	3	2	1
28	27	26	25	24	23	22	21	20	19	18	17	16	15
42	41	40	39	38	37	36	35	34	33	32	31	30	29
56	55	54	53	52	51	50	49	48	47	46	45	44	43
69	68	67	66	65	64	63	62	61	60	59	58	57	
82	81	80	79	78	77	76	75	74	73	72	71	70	
95	94	93	92	91	90	89	88	87	86	85	84	83	
108	107	106	105	104	103	102	101	100	99	98	97	96	
121	120	119	118	117	116	115	114	113	112	111	110	109	
134	133	132	131	130	129	128	127	126	125	124	123	122	
148	147	146	145	144	143	142	141	140	139	138	137	136	135
162	161	160	159	158	157	156	155	154	153	152	151	150	149
176	175	174	173	172	171	170	169	168	167	166	165	164	163
190	189	188	187	186	185	184	183	182	181	180	179	178	177

24 Evaluation of Specialized Literature

The schematic color perception space, gauge for the subject color in specialized literature

There is a great deal of specialized literature in which color plays in important role. Many authors still use the outdated color circle or an outdated three-dimensional color diagram. The complementary color pairs do not lie diametrically opposite to each other in the old color circles; the basic colors of cyan and magenta, as well as the colors derived from them, are usually missing. The colors are not ordered as to their own inherent lightness in relationship to the lightness values of the lightness axis in outdated three-dimensional color diagrams.

Valuable information

Such books, however, sometimes offer important information as far as a given subject is concerned. They can not be used as a standard for mutual color relationships; we can superimpose our own knowledge of mutual color relationships within the new color perception space. We can 'correct' complementary values and mutual color and lightness relationships where needed. The assertions based on outdated theories can be transformed into usable data. Exercises and assignments for professional education in color can be changed, by a correct approach, into useful tasks.

Implementation of the new color order

It may seem difficult to start using a new color order for someone who has long been accustomed to the old methods, especially where one's profession is concerned. It is, however, much easier than it seems. The advantage of the new method is that one is up to date and can vouch for one's work. An unnecessary limitation of the color palette results from outdated color ordering. A limitation of possibilities does not necessarily lead to poor results, on the contrary; but the opposite is also not true, a random limitation of the color palette certainly does not guarantee good results.

No random limitation of the color palette

The people of what was formerly the Congo, knew how to create decorative and harmonious color combinations with a limited color palette. Their colors consisted of ocher yellow, red-brown, black, white and sometimes ivory, plus a spot of green or a chalky light blue.

Even so, the designer of a spring collection of women's fashions could certainly wish to keep other colors in mind.

Professions with color aspects

There are many professional areas in which the 'aspect of color' is important such as: painting, graphic art, photo and film, glaze and enamel art, theater and ballet, textile art, fashion, interior design, architecture, garden architecture, lighting engineering (color signals and color coding systems for traffic management and other systems), color consultations, color study, psychology. The information about color hue, lightness and saturation are only a few of the many color aspects which are treated in the textbooks for the diverse professions. Color influence factors can be quite different in the various professions. Even though there are many differences in color properties, all colors can be described with three qualities: color hue, lightness and saturation. It is therefore important to

---

Captions for the illustrations on page 78

#137 Old color circle; the color perception schemes show irregularities.

#138 New color circle; the color perception schemes are regular.

Subtractive color formation: deducting
#139
WHITE	- RED	= CYAN
CYAN	- GREEN	= BLUE
BLUE	- BLUE	= BLACK

#140
WHITE	- BLUE	= YELLOW
YELLOW	- RED	= GREEN
GREEN	- GREEN	= BLACK

#141
WHITE	- GREEN	= MAGENTA
MAGENTA	- BLUE	= RED
RED	- RED	= BLACK

Additive color formation: adding
#142
GREEN	+ BLUE	= CYAN
BLUE	+ RED	= MAGENTA
RED	+ GREEN	= YELLOW

#143
CYAN	+ RED	= WHITE
MAGENTA	+ GREEN	= WHITE
YELLOW	+ BLUE	= WHITE

Partitive (optical) color formation: averaging
#144
| BLUE/BLACK | + GREEN/BLACK | = CYAN/BLACK |

#145
| GREEN/BLACK | + RED/BLACK | = YELLOW/BLACK |

#146
| RED/BLACK | + BLUE/BLACK | = MAGENTA/BLACK |

Note approaches with new insights

test all theories as to the laws of our color perception possibilities. It is not difficult to learn to think in terms of the complementary color pairs: ultramarine blue/yellow, green/magenta, and red/cyan. Assignments in professional books which are based on the old color circle, can easily be transformed into similar assignments according to our new knowledge of complementary color pairs, lightness relationships and saturation relationships. All valuable professional information, as far as color is concerned, can only become of more value. The old books have not become worthless because of the new theories, on the contrary, they will be of more use when supplemented by our own insight into the laws of our color perception possibilities.

25 Build Your Own Model of the Color Perception Space

Closed form

Primarily the outer surface

Much attention is focused on the closed outer shape of the illustrated color models. We see the cone, double cone, pyramid, double pyramid, ball, cube, elongated cube and cylinder. We see a schematic presentation of the outer color surfaces in these closed forms. We can not see how the colors change towards the center of the color model.

Open form

Color progression inside the color space

It is much easier to see the progression from full color to colorless in a color model with an open form. It is also possible to see the color progression of any randomly chosen line inside the color space.

There are many ways to create an open color model. The possibilities fall into two groups: the linear form built of spokes, wires i.e. or a form built up of surfaces (#88).

The open form, built of surfaces

There are many advantages to building the model from surfaces, especially if these are placed vertically as a fan around the lightness axis. It is possible to have an overview, in this type of arrangement of color surfaces, of one section of lightness gradations and of saturation gradations per color hue. If we use transparent material, the whole color perception space will be easier to 'read'. We must find a way to construct this fan-shaped model so that each color hue surface can be taken out of the model separately.

Transparent

We can, for example, choose a base of thick material with a sufficient number of deep grooves (with a cylindrical central 'axis') to accommodate the transparent color hue plates and to allow them to be slid in and out of the base.

Glass or PVC

We can use beveled glass, or transparent pvc (plexiglas) for the color hue panels. Glass is less expensive and there is less risk of glue spotting but it can break easily.

PVC is more expensive, is easier to glue-spot, not as breakable and easier to work with. Make the model as large as is practically possible; do remember that it will need a place for safe-keeping. Larger dimensions do not make construction more difficult, and the result is easier to 'read'. The costs of a larger model, of course, must be considered; pvc is especially expensive.

Determining the size

Illustrations #125 -#136 show 12, schematically in color, color hue gradations. Determine how high your model will be: l00 cm or more, 75 cm, 50 cm or smaller. Calculate about 10% more for the part of each plate which fits into the grooves of the base (#88). Determine the diameter in proportion to the height. Make a drawing to the actual size of a glass (pvc) plate and draw a line across the bottom to indicate where the plate fits into the base. Draw half circles along the right edge to correspond with the edge of the plate. Place the top and bottom half circles a little from the top edge and bottom line. Divide the lightness axis from white to black as in #89A in equal steps as, for example, in A -J #118.

One drawing

Draw the half circles A to J and place a thin line horizontally from the middle of each half circle to the other (left) side of the plate drawing. Divide these horizontal lines in equal saturation steps. Draw the round spots with a circle stencil, as you have indicated, and to the same scale as the half circles (#89B). It is now possible to paste all spots on all the glass plates correctly, with the aid of this one drawing, by placing each plate individually onto the drawing. Use a punch of the size you have chosen for the spots (there are punches up to at least 5cm) to punch them out. It is best to use beechwood or gray punch-board as a punch base. Color each spot on two sides. It may be easier to paint a piece of paper on two sides from full color to gray of equal lightness; punch dots from this paper and choose the gradations from a wide range. First choose a spot with a color halfway between full color and gray, then between gray and the halfway color etc. Paste the spots onto the glass with transparent glue which is suited to paper and glass (actually paint and glass) or paint and pvc. Remove any extra glue with a clean dry cloth.

Color on both sides

Mat or glossy color spots

We now have a glass plate with mat paint colors on one side and glossy colors on the other side as seen through the glass. Start with the gray gradations between black and white (#118), followed by the six basic colors (#117, center); next, choose all full colors to white and all full colors to black (#123 & 125); finally, all full colors to gray of the same lightness as the intrinsic lightness of the given full color(s) (#123 & 125). When we determine the size of the spots, as related to the size of the whole model, we must also calculate the horizontal spot pattern. The total number of color hue surfaces must be divisible by two and three in order to achieve a mathematical circumference distribution of the color cylinder. It must be divisible by two in order to be able to place the complementary color pairs opposite to each other; and by three in order to be able to place the three eye primaries at mutually equal distances from each other. There are 12 schematically presented color hue surfaces in illustrations #125 -136. If we slide all the glass plates into their grooves in the base, we have constructed a fan of color hue surfaces in which the colors are ordered as to lightness and saturation. It is easy to take (slide) a whole color hue surface, or two which lie opposite to each other, out of the color perception space model. We can view each complementary color pair on a flat surface and see how each color progresses from full color to neutral and back to full color. It is very instructive to see all these color changes.

Main divisions divisible by 2 and 3

Main line of ordering phases per (historical) period:

600 B.C. to Newton (1660)	Ordering from one lightest color to one darkest color from white to black.
Newton (1660) to Munsell (1915)	No relationship between inherent lightness of a color hue and the corresponding lightness on the neutral axis, except for Titchener, a first attempt (1887).
Munsell (1915) to Hesselgren (1955)	Ordering from one lightest to one darkest color, in relationship to the lightness gradations of the neutral axis.
Maxwell, CIE, Hickethier cube and rhombohedron (1857-1972)	Three light and three dark basic colors, inherent lightness of the color hues does not agree with the lightness gradations of the neutral axis: three-color printing, regular zigzag line, visually irregular color steps from light to dark.
Gerritsen, from 1975	The schematic three-dimensional color perception diagram; all colors ordered according to color hue, inherent lightness and saturation; irregular zigzag line (color perception); visually regular color steps, also as to light and dark values.
CIELAB/CIELUV (#119) and the present-day Munsell system	CIELAB and CIELUV (see opponent values, #119), as well as the present-day Munsell system, have development possibilities within their organizations for adapting to the new ordering of the Gerritsen color perception space.

26 Emotional Color Values

Color or non-color

We now return to the most important aspects of color, as we did at the beginning of this book. What influence does color have on us, and under what circumstances? What message do we want to put across? What atmosphere do certain assignments require? Can color express what we want to say or perhaps we should consciously choose a neutral in whites, grays or blacks? Shall we choose a neutral with more emphasis on the expression of the kind of material? No one can give us a recipe as the answer.

Which atmosphere?

We can discuss our assignment with colleagues, we can listen to many points of view. Beyond this we must let ourselves be led by one principle: 'What atmosphere does our assignment require and how can we best 'express' this at this time under these circumstances for the 'audience'?' We must, ultimately, make choices. Is this our own choice or a spark we have received because we remain willing to receive...?

I have heard that designing is making new use of, or giving new definition to, existing information. I remember vividly a color pink-salmon-rose which I would recognize from thousands of colors. It is the color of a metal doll's bed that my younger sister received from Santa Claus himself many years ago. The bed was about 15 cm long, it was lined with downy pillows and it smelled of soap...

The same color—a different situation

Sometimes we recognize exactly the same color in a totally different situations and very different values! I saw the same pink-salmon-rose as the solid color of a sky in a pastel drawing of the Provence. Another time, I

Other expression values

saw the color as the shimmer on a copper kettle on an oil painting. I must have seen the color as color sample in the Munsell Atlas, I saw the color between make-up and nail polish colors and at the sea shore in a shell...

The same rose color expressed very different things in each different situation. Our sub-conscious is filled with a treasury of memories of color recognitions, color expressions and of color harmonies. We know colors of nature, of sky, water and land, plant and animal, through all seasons, all over the world. All these colors and color combinations will activate different emotions in different situations. The same is true for mixed colors we may have seen originate on our palette, for colors in advertising, colors of graphic art and paintings, colors of jewelry, colors of spouting fountains, neon advertisements, activated phosphors, metallic colors etc. We think of all colors in the world which man has built himself. There is an inexhaustable source of color associations; we do not, however, yet have suitable advice on how to use all this color.

Color harmony theories

Complementary contrast is not a requirement for color harmony

Total color image

Many attempts have been made, through many historical periods, to create recipes for color harmony. It is, however, not possible to make a list of rules to describe the harmonious or disharmonious visual image. Complementary contrast, whatever the subject, is not a requirement for a harmonious color image. 'Ton-sur-ton' (colors all related to one color hue in slightly different shades or tints) color use doesn't guarantee harmony either. The harmony theory of one, two, three and four tone/hue accords is also not a harmony requirement. An interplay between mutual color surface sizes is most certainly necessary. The total impression, also in three dimensions, must be kept in mind. The size scale of any given object is an important influence with respect the expressive impact of color combinations. The most important requirement is: 'Is there harmony and unity in the total of the chosen color scheme and the atmosphere inherent to the given subject as to its own distinguishability and/or application.'

The development of the color sense

From "Color"

We can learn to develop our sense of color harmony as to the atmosphere of a given subject. I wrote a section, in my book Theory and Practice of Color, 'The development of color sense' (in the chapter called 'Training'): 'We can develop color sense through observing and analyzing colorful objects, and through experimenting with paint colors. Observable objects with the greatest color harmony can be found in nature. Our brains translate the functional absorption and reflection of the sun's light energy by air, earth, water, plant and animal, via our visual system, into a harmonious visual scene. Everything in nature is tuned to the rest and has a function in the totality. The amount of absorbed and reflected sun energy, (which we experience as color) necessary for good functioning, is in agreement with the harmony of the life rhythm of living nature. We observe colors of air and earth as a whole, for example in a landscape. We observe the colors of plants and flowers as a whole, in their harmonious surroundings. Sometimes we take a flower from its natural surroundings and look at its color at home. We look at the whole and at its mutual color relationships. We also look at the separate parts which make up the mutual color relationships. We can study the colors of an animal or person. Our color association sensitivity values can also be developed by the conscious study of all this beauty in form and color harmony.'

End of quotation from Theory and Practice of Color. The harmony and beauty values described here follow the laws and rhythm in harmony with the infinite...

Naturally, we can not simply take the example of the proportions in form and color of a beetle and use these as a color scheme for a living room. A living room is not a beetle and certainly not this one particular beetle. We have developed a sense of value for form and color harmonies. It is possible for us to consciously choose our color scheme if we bear in mind the indivisible essence of harmony between form and color. Our choices remain, however, bound by universal values within our time, place and cultural traditions.

Developing a value sense for color harmonies

27 Professional Pointers

Guidelines for each special field

Special instructions as to color use in the various professions can be found in specialized literature.

There are special rules, for some professional color usage, which must be followed to achieve successful results, along with attention to sphere of the project goal. The artist will take the type of light he uses into account. He will choose the color palette he thinks he needs for a given project. Which red he chooses is important to him. He will choose with care the materials he thinks are of the best quality for his work. He will take into account the mixing results of this red with other chosen colors. He already knows the typical differences in the mixed color of a certain white with the colors thallium-blue, cobalt-blue, ultra-marine blue or prussian-blue. He will not mix his own vivid orange from yellow and red. Cadmium orange is much more pure of color, he can change this pure color a little, if necessary, with some yellow or red. He does not mix a red from magenta and yellow. He chooses a red straight from the manufacturer's paint tube because it is much more pure. He can always change the red slightly, to his liking, with a smidgen of carmine or orange. If he wants magenta or cyan, they must come straight from a tube. He can always mix a little to get the exact color he wants.

Skill and experience

Pure colors, straight from the tube

Why blue mixed with yellow usually produces green

Most blues let green through, light wavelengths of the middle area of the daylight spectrum. These blues, mixed with yellow, result in a green and not a neutral gray as in partitive color mixing. These colors show a much more favorable mixing result than magenta and yellow to yield red. This is most true for opaque paints. The artist will, even so, take green from the tube if he wants a brilliant color, and mix it to change the hue just a little. This book was not written as a guide for paint mixing; there is enough information available as to the paint mixing ideas according to the old color circle. It was written to indicate how color surfaces which lie next to each other influence each other; visually, these influences of one color

Blue and yellow, green or gray?

Color influenced by adjoining color surfaces

upon another or mutual color influencing, occur according to the laws of our color perception. The artist always works with this influence which colors have upon each other. It is actually the secret of the origination of the colors which he shows us.

He first chooses a certain technique and adds his own personal touches. He can best express himself for a given subject in pastel colors/technique, or water color, gouache, oil or acrylic paint, and further with palette knife, brush or pencil.

The graphic artist thinks in other materials, tools and in reproduction techniques.

The photographer or film artist chooses his camera and lenses, brand and type of film. He experiments with very sharp or diffuse focus, under- or over-exposure, influencing color with various filters during exposure or in one of the different development stages.

Properties, scenery colors, costumes and lighting must be coordinated for theater and ballet.

The interior decorator works between clarity and disguise. His work should create a total atmosphere which is experienced as recognition and guidance without the explicit 'presence' of the decorator. He must take into account the light coming in through the different windows. He works with the north, east south or west light as well as the influence of light from large reflecting surfaces in the neighborhood. These could be color remissions from living nature or from the color of buildings in the area.

The garden architect and the artist in flower arrangements will take into account the characteristics of the flowers and plants. They work with effects of single items and also with the total effects of groups. These are very different concepts. This is true for the general structure, form, mutual properties and for color.

Let each person study general color relationships and those pertaining especially to his profession and make sure his ideas are up to date. Is the essence of the objective of primary importance? Does the new creation have style inherent to the original assignment? We can find worth while advice, in this publication, to help us translate visual values of the language of color for our use; and to lead us to answers to the questions above. We can understand different ideas about colors and their mutual relationships, in one long line from past to present, which will give us a good background for better insight into this complicated material. Philosophies, concepts, physical and chemical experiments have led us to scientific study as to the laws of our color perception possibilities. We learn to know the seemingly illogical behavior of subtractive, additive, and partitive color mixing (#139-146) as normal laws which concern our color vision.

We know that a great lightness contrast prevails above a color contrast of equal lightness. We have studied the inherent lightness of full colors. We know which colors are relatively light saturated colors and which are relatively dark saturated colors. The visual image is determined mainly by lightness information, enriched by additional color information.

We have been made aware of how important the kind of light is with which we see, because it is the source of information at a distance which delivers energy to our eye in the process of sight. Wavelength areas which are not present in the light source, can not be activated, and therefore the perception of the colors not present in the light source is not possible as it is in normal daylight. Think of color renditions under sodium lamps and uncorrected mercury lamps. The artist know the importance of constant quality daylight in his studio and chooses a window with northern

Other means to create a chosen sphere

Every field has its own aspects

Type of light—direct or indirect

The essence of the objective is of crucial importance

Artists and perception laws

Subtractive-additive-partitive

Light is the only source of information

Artists and perception laws

Light is the only source of information

exposure. Artists who work with color have, for centuries, searched for the greatest visual color contrasts. They thought to have found the answers in the colors red, yellow and blue and the presumed contrasting colors green, violet and orange.

The study as to the laws of our color perception have proved to be otherwise. The laws of our color perception also function for the artist painter with normal vision. The complementary basic color contrast pairs are: ultra-marine blue/yellow, green/magenta and red/cyan. Between these basic colors lie an infinite number of complementary color pairs. The greatest light-dark contrast is black-white.

We so often hear, 'But for the painter the colors are red, yellow and blue'; therefore I have wished to emphasize the incorrectness of this statement that red, yellow and blue are 'the' colors. I have gone into much more detail, as to color behavior and practical applications, in my book Theory and Practice of Color (the new color study based on the laws of color perception). I hope that this book, Evolution in color, has made a contribution to better understanding of color and mutual lightness and color relationships; and how this color ordering can be used as a guideline for the theory of color.

We wish to express our heartfelt thanks to all those who in one way or another, for content or design, contributed to this publication